The NASA STI Program Office . . . in Profile

Since its founding, NASA has been dedicated to the advancement of aeronautics and space science. The NASA Scientific and Technical Information (STI) Program Office plays a key part in helping NASA maintain this important role.

The NASA STI Program Office is operated by Langley Research Center, the lead center for NASA's scientific and technical information.   The NASA STI Program Office provides access to the NASA STI Database, the largest collection of aeronautical and space science STI in the world. The Program Office is also NASA's institutional mechanism for disseminating the results of its research and development activities. These results are published by NASA in the NASA STI Report Series, which includes the following report types:

- TECHNICAL PUBLICATION. Reports of completed research or a major significant phase of research that present the results of NASA programs and include extensive data or theoretical analysis. Includes compilations of significant scientific and technical data and information deemed to be of continuing reference value. NASA's counterpart of peer-reviewed formal professional papers but has less stringent limitations on manuscript length and extent of graphic presentations.

- TECHNICAL MEMORANDUM. Scientific and technical findings that are preliminary or of specialized interest, e.g., quick release reports, working papers, and bibliographies that contain minimal annotation. Does not contain extensive analysis.

- CONTRACTOR REPORT. Scientific and technical findings by NASA-sponsored contractors and grantees.

- CONFERENCE PUBLICATION. Collected papers from scientific and technical conferences, symposia, seminars, or other meetings sponsored or cosponsored by NASA.

- SPECIAL PUBLICATION. Scientific, technical, or historical information from NASA programs, projects, and mission, often concerned with subjects having substantial public interest.

- TECHNICAL TRANSLATION. English-language translations of foreign scientific and technical material pertinent to NASA's mission.

Specialized services that complement the STI Program Office's diverse offerings include creating custom thesauri, building customized databases, organizing and publishing research results . . . even providing videos.

For more information about the NASA STI Program Office, see the following:

- Access the NASA STI Program Home Page at *http://www.sti.nasa.gov*

- E-mail your question via the Internet to help@sti.nasa.gov

- Fax your question to the NASA STI Help Desk at (301) 621-0134

- Telephone the NASA STI Help Desk at (301) 621-0390

- Write to:

  NASA STI Help Desk
  NASA Center for AeroSpace Information
  7121 Standard Drive
  Hanover, MD 21076-1320

NASA/TM—2004-211522

**EMERGING COMMUNICATION TECHNOLOGIES
(ECT) PHASE 2 REPORT
Volume 3
ULTRA WIDEBAND (UWB) TECHNOLOGY**

Gary L. Bastin, Ph.D.
ASRC Aerospace Corporation, John F. Kennedy Space Center, Florida

William G. Harris, PE
ASRC Aerospace Corporation, John F. Kennedy Space Center, Florida

Robert Chiodini
ASRC Aerospace Corporation, John F. Kennedy Space Center, Florida

Richard A. Nelson
NASA, YA-D7, John F. Kennedy Space Center, Florida

PoTien Huang
NASA, YA-D5, John F. Kennedy Space Center, Florida

David A. Kruhm
NASA, YA-D5, John F. Kennedy Space Center, Florida

**National Aeronautics and
Space Administration**

John F. Kennedy Space Center, Florida 32899-0001

September 2003

Acknowledgments

Although there is always the risk of inadvertently forgetting someone, the ECT team nonetheless wishes to acknowledge especially the assistance and guidance provided by the following individuals, listed alphabetically. Without the continued support of these supporters who believed in the value of this project, this project could not have accomplished all its goals.

Name	Organization
Hugo Delgado	NASA-KSC
Eric Denson	NASA-KSC
Temel Erdogan	Dynacs
Mike Grant	CSR-Tel-4
Debra Holiday	FL Space Authority
Don Hoover	CSR-Optics
Gary Janousek	CSR-XY
Chris Kerios	Dynacs
Ray Knighton	ITT-SLRSC
Dennis McCunnion	CSR-TVOC
Jules McNeff	NASA-HQ
Rich Nelson	NASA-KSC
Don Philp	Dynacs
John Rush	NASA-HQ
Jim Shaver	NASA-Hanger AE
Steve Schaefer	Dynacs
Steve Schindler	NASA-KSC
Darin Skelly	NASA-KSC
Stan Starr	Dynacs
Dave Struba	NASA-HQ
Lisa Valencia	NASA-KSC
John Walker	CSR-JDMTA
Phil Weber	NASA-KSC

Available from:

NASA Center for AeroSpace Information
7121 Standard Drive
Hanover, MD 21076-1320

National Technical Information Service
5285 Port Royal Road
Springfield, VA 22161

TABLE OF CONTENTS

No.	Description	Page

Executive Summary

The National Aeronautics and Space Administration (NASA) is investigating alternative technologies to facilitate building communication networks for future Spaceports and Ranges. This reports documents an investigation conducted from October 2002 through September 2003 of an emerging communication technology known as Ultra Wideband (UWB) communication. Contained in this report is an overview of UWB communication technology, a survey of UWB equipment vendors, and complete details of the theoretical and experimental research that was performed during this emerging communication technology investigation.

The summary conclusion of this report is that UWB communication holds great promise for augmenting future Spaceport and Range communication networks through enhancing short-range, high speed, wireless communication. This enhancement is accomplished through simultaneously integrating *position-aware* functions with *traditional communication* functions. UWB technology achieves this dual-function integration through using short impulses instead of the continuous waveforms common to most wireless systems. Because short impulses are used, UWB communication links are inherently immune to most multi-path interference, and also achieve better instantaneous spectrum re-use among users. UWB modulation also provides a Low-Probability-of-Detection (LPD) waveform with selectable security. UWB systems can therefore provide fade resistant, high speed data links wherever the presence of easily detectable wireless transmissions must be avoided, such as in specialized tactical situations.

Despite the many theoretical advantages that exist for UWB modulation, much anxiety commonly arises with UWB emissions because of their ultra wide bandwidths. This characteristic especially causes concern *vis-à-vis* possible deleterious effects to narrowband legacy systems that typically operate at low link margins, such as GPS navigation systems, which often operate with link margins of only 1 or 2 dB. Though this concern has been addressed previously in Federal regulations through the inclusion of spectral emission mask requirements for UWB emissions, not all of the technical concerns have been settled. This report further researches some of these interference concerns by investigating UWB interference to, and UWB susceptibilities from, legacy wireless systems.

The summary recommendation of this report is that UWB technology appears to hold many of the key advantages needed to tackle a number of wireless systems requirements needed in the future, provided it can co-exist successfully with both legacy narrowband wireless systems, and with other UWB wireless systems. Based on the demonstrated performance seen during testing, there is ample reason to believe that UWB wireless systems *will* be able to coexist with most legacy wireless systems. This report is a first step in understanding and assessing UWB's applicability for supporting the communication needs of future Spaceports and Ranges.

1.0   **INTRODUCTION**

Ultra Wideband (UWB) wireless technology is the prime candidate for becoming the next step in the evolution of wireless technology. It is potentially well suited for use wherever high-speed data rates (to at least several hundred Mb/s) are desired over ranges up to several hundred meters in locations prone to fading due to multi-path propagation. This emerging wireless technology uses short duration pulses known as *monocycles* to propagate signals over physical distances instead of the sinusoidal carriers used by legacy wireless systems.[1]

Two major UWB wireless technology application areas exist today, addressing communications and radar needs, respectively. This report largely focuses on UWB communication applications since UWB radar applications will likely not see widespread use within the communication networks of future Spaceport and Range.[2]

Whether occupied all in one band, or sub-banded into 5 to 15 sub-bands, fundamental UWB communication concepts in use today all derive from simpler pulse-based systems first used in radar systems. The modulation waveforms currently used in UWB systems today have not changed significantly since their first use over 30 years ago in radar systems. As a result, UWB wireless systems often retain many traditional radar capabilities, even when intended solely for communication purposes.

This characteristic capability of UWB technology is expressed by stating that UWB systems are *position-aware;* that is, receiving UWB modulated signals requires an inherent, automatic assessment of relative distances among the transmitters and receivers within a UWB wireless network. Coupling communications with position-aware features simultaneously enables wireless systems based on UWB to provide capabilities that were never previously possible in traditional wireless communication systems.

In spite of occupying very large bandwidths, UWB is often found to be extremely benign to existing wireless systems and services. The use of ultra wide bandwidths also has advantages relative to narrower bandwidths. Since the correlation bandwidth of the dense urban and dense structure propagation channel is typically less than 10 MHz over 3.1 GHz to 10.6 GHz, the use of extremely short-duration bursts achieves ultra wideband occupancy over much greater than the correlation bandwidth of the channel and this completely mitigates the effects of destructive interference (i.e., fading) in multi-path signals.[3] Because of this advantage, a high fidelity UWB replacement for FM tactical radios would completely avoid much of the fading so commonly heard when operating in

---

[1] Moe Z. Win and Robert A. Scholtz, "Ultra-wide Bandwidth Time-Hopping Spread-Spectrum Impulse Radio for Wireless Multiple-Access Communications", IEEE Trans. Comm. Vol. 48, No. 4, April 2000.
[2] UWB radar functions will still likely play a critical role in enhancing *security* around future Spaceports and Ranges; they just will not play any significant role within the *communication networks*.
[3] Correlation bandwidth refers to the bandwidth over which a spectral null is typically correlated and all frequencies fade simultaneously. It is the bandwidth over which a fade exists in, for example, an urban channel. Any signal within this bandwidth is simultaneously lost during fading events, and the fade is said to be 'correlated' over this range of frequencies.

dense urban downtown areas and other dense structure areas, such as within many office buildings. This is a key advantage for tactical radios based on UWB technology.

The numerous capabilities engendered by UWB technology, investigated on a purely technical basis rather than on an economic or political basis, are especially intriguing. UWB largely renders data compression technology obsolete. The requirement to pack more and more bits into a *limited* bandwidth is largely eliminated with UWB since the bandwidth can be selected to be arbitrarily wide with UWB technology. In addition to the purely technical performance advantages of UWB technology, UWB also has the inherent economic advantages typical of a disruptive technology. UWB transmitters and receivers do not require all of the oscillators, mixers, filters, and numerous other expensive radio frequency (RF) components required in conventional wireless gear. As discussed earlier, UWB likewise eliminates data compression and de-compression chip-sets, as well as eliminating the dc power required to run these data compression/de-compression chips.

The end result is that UWB equipment often requires lower-cost components totaling only around ten percent of the cost of the components required to implement conventional wireless gear. Likewise, UWB gear can use batteries that are only 10% to 25% of the cost, size, and weight of batteries required for existing wireless battery-powered equipment due to improved power efficiencies of the short-duration transmitted signals, elimination of data compression, and elimination of other power-consuming functional blocks.

Because of these economic and performance advantages, UWB communication gear has considerable advantages over existing wireless gear. UWB systems can provide:

- Voice and data communication with selectable degrees of security
- Indoor, through-the-wall, and perimeter security radar functions
- Precise ranging capability to determine the precise distances between objects with real-time tracking to within an inch
- Elimination of data compression requirements to fit data into pre-set narrow bands
- Nearly complete immunity to multi-path propagation, such as encountered in dense, urban areas, simultaneously increasing data throughput as well as avoiding low signal levels due to destructive interference (fading) of received multi-path signals

With these diverse capabilities, UWB technology can enhance numerous Spaceport and Range disciplines including:

- Wideband operation during a launch event, in spite of considerable multi-path reflections caused by aluminum-based particle exhausts
- Real-time tracking of high cost assets, with high precision
- Reliable, high-speed, secure wireless voice, data and video transmissions inside buildings
- Personal radar for security system functions for perimeter control
- Radar functions, with through-the-wall sensing to penetrate materials such as brick and concrete to provide more defined images than conventional radar for security sweeps of buildings and cargo areas of tractor trailers

SBIR investigations of UWB technology have also been conducted in coordination with Johnson Space Center to enable in-helmet video transmission in next generation spacesuits.

In short, UWB represents a major shift in terms of implementation capabilities. Further, because of battery life extensions, it is possible to tailor the battery-life to reduce the cost of existing batteries through eliminating materials. With all the benefits, as well as the cost reductions possible, UWB technology is truly a disruptive technology worthy of consideration for use on future Spaceports and Ranges, especially for short distance communications.

1.1   UWB REGULATORY AND TECHNOLOGY OVERVIEW

1.1.1   Regulatory Overview

Current UWB applications typically use one of two fundamental types of modulations: Time-Hopping (TH) Pulse Position Modulation (PPM) or Bi-phase Pulse Modulation. By current FCC Part 15 rules adopted February 14, 2002, a total of 7500 MHz of unlicensed spectrum is available for UWB communication over 3.1 to 10.6 GHz.[4] The present UWB communication rules specify neither the exact modulation or waveform shapes that must be used; instead, only the maximum effective isotropic radiated power (EIRP) levels (-41.3 dBm/MHz), the maximum permitted frequency spectrum allocation (3.1 GHz to 10.6 GHz, for emissions above a maximum spectral mask limit of 10 dB down from the peak radiated emission of the complete system, including the antenna), and additional usage specifications (indoors, ac power only) are established. This *laissez faire* approach sets the minimum characteristics necessary to encourage the peaceful co-existence of

---

[4] See: 47 CFR Ch. I, Part 15, Subpart F Ultra-Wideband Operation, (10-1-02 Edition). Available from: http://www.gpoaccess.gov/fr/index.html (Retrieved 21 August 2003.)

UWB transmissions among more established narrowband transmissions, while still permitting UWB innovation to continue largely unhindered.[5]

Because of this legislated freedom, there are at present two approaches used for occupying the allocated 7,500 MHz of unlicensed spectrum. So-called *old UWB* equipment occupies as much of the 7500 MHz bandwidth simultaneously as the electronics and antenna can actually accommodate. In practice, typical bandwidths still span only 2,000 MHz to 4,000 MHz out of the total 7,500 MHz that is permitted when implementing UWB communications using commonly available (and low cost) semiconductor processes.

Reconciliation of the limitations of affordable semiconductor process implementations of UWB communication ICs (integrated circuits), with only a partially filled one-band spectral occupancy, has led to newer proposals, set forth during 2003 at IEEE 802.15.3a standards Task Group 3A (TG3a) meetings to improve UWB spectral efficiency. These proposals recognize the inability of current generation low-cost hardware to occupy 7500 MHz of bandwidth simultaneously by instead dividing this UWB spectrum into multiple sub-bands. This sub-banded approach is now being called *new UWB* by several vendors.[6] Various numbers of sub-bands are proposed for meeting the proposed 802.15.3a specifications, ranging from 5 sub-bands up to 15 sub-bands.[7]

Regardless of the exact number of sub-bands ultimately selected, there are many advantages to sub-banding the allocated UWB spectrum. The semiconductor processes that can supply less-expensive solutions, usable only over the lower sub-bands (e.g., CMOS or SiGe), can still be used. Then, as semiconductor-processing technology improves and/or processing costs drop for higher performance processes, the higher sub-bands can subsequently be occupied. Likewise, specific sub-bands that may cause interference in particular locations can simply be turned OFF in *new UWB*. For example, spectrum in and around 5.5 GHz, falling in sub-band 2 of *new UWB*, is also used by recently introduced IEEE 802.11a standard wireless Ethernet (Wi-Fi) hardware that runs at 54 Mb/s. For locations where this 5.5 GHz spectrum is occupied by 54 Mb/s Wi-Fi hardware, the newer sub-banded UWB approach would elegantly allow simply avoiding sub-band 2, thereby improving the peaceful coexistence of UWB among narrowband wireless legacy systems. An additional advantage would be the possibility of running multiple (i.e., perhaps up to 4 or 5, or possibly even up to 14 or 15) *piconets* in the same local area through utilizing a different UWB sub-band for each *piconet*.

---

[5] Unfortunately, as of early August 2003, this inexactness has led to rogue proposals for implementing the IEEE 802.15.3a standard for which not all are truly UWB transmissions. Instead, in order to occupy the necessary bandwidth to be classified legally as UWB, some proposals, using more narrow-band modulation schemes, have merely included pilot tones to occupy enough bandwidth to achieve classification (technically) as UWB transmissions and which accomplish little else, adding no performance enhancements.

[6] No doubt a different moniker will arise shortly in place of *new UWB*, as even newer UWB advances occur.

[7] As of the writing of this report (August 2003), no resolution of the number of sub-bands ranging from 1 to 5 to 15 has occurred.

Among the major companies, there is still no consensus on how best to provide IEEE 802.15.3a implementations that utilize the Part 15 allocated bandwidth, whether through sub-banding, or through using but one band. In late July 2003, fifteen of the major UWB companies combined their approaches and merged the Intel-led multi-band approach with the Texas Instruments' led multi-band approach through settling on one common multi-band approach and establishing the Multiband-OFDM (Orthogonal Frequency Division Multiplexing) Alliance (MBOA). The major members of the MBOA include Texas Instruments (TI), Staccato (formerly Discrete Time), General Atomics, Time-Domain, Intel, Panasonic, Mitsubishi, Philips, and Samsung. Still proposing a single-band approach, at odds with the approach proposed by the MBOA, are XtremeSpectrum, Motorola, STMicroelectronics, Communications Research Lab, the University of Minnesota, and ParthusCeva. At least two of the single-band proponents, XtremeSpectrum and STMicroelectronics, are proposing CDMA (Code Division Multiple Access) in addition to Bi-Phase Pulse modulation.[8]

A series of meetings were held by the FCC in early August 2003 to collect information on the two opposing camp's viewpoints in an attempt to reach consensus on the best implementation to endorse for occupying the 7500 MHz of allocated Part 15 UWB bandwidth. At the present time (i.e., late August 2003 through early September 2003), no final decision has been made by the FCC as to which proposal to endorse.[9] Until a formal decision is made, reaching an industry-wide consensus for standardizing UWB communication links for WPAN/WLAN applications similar to Wi-Fi will likely not occur. Because of this, most UWB chipset developments have been placed on hold, awaiting a final FCC decision.

1.1.2   Technology Overview

UWB communication systems use very low power (Part 15 levels are 5 mW or less), unlicensed, very short duration (< 2 ns, typically 10 to 1000 ps) UWB pulses at repetition rates from 10 to 40 MHz. Centered at a typical center frequency of 2 GHz, first-generation UWB typical system occupied 1.4 GHz. To avoid interfering with GPS signals and other low-power signals below 2 GHz, newer UWB systems, in compliance with current Part 15 UWB requirements, now occupy 3.1 to 10.6 GHz, either in one band, or within several sub-bands.

Because the pulses are pseudo-randomly (PN) shifted in time, transmitted signals resemble white noise to narrowband, conventional receivers. Because of their wideband, low-power characteristic, UWB systems typically co-exist with existing narrowband communication systems, without causing significant interference. Likewise, because of their high processing gains of 30 dB or better due to occupying wide bandwidths, noise rejection performance of UWB systems is superior to that seen in narrowband systems.

---

[8] Outside Plant Magazine, August 7, 2003, http://www.ospmag.com/op_enl/inside_scoop.htm, retrieved 25 August 2003.
[9] Patrick Mannion and Robert Keenan, *"Samsung taps Staccato for wireless personal nets,"* Electronic Engineering Times, August 18, 2003.

Since the short duration pulses provide excellent multi-path immunity, the pronounced fades seen within buildings, or around a launch pad, with conventional narrowband systems are avoided. The use of short pulses enhances communication reliability of wireless LANs and other systems using UWB technology. In addition, because of the precise timing inherent from the time-modulated characteristics, precise position location functions are inherently features of UWB.

For a given range, limited mostly by peak powers, UWB systems provide an especially attractive solution for portable, battery-powered applications. Because they employ pulses, the average power is extremely low (5 mW, or less), whereas the range associated with the systems is more like that seen for transmitter powers of 30 dB or so higher, as associated with their peak transmitter powers. In other words, a 5 mW average power signal is equal to 6.98 dBm; a peak power of 30 dB higher is equal to 36.98 dBm, or, in terms of Watts, 5 Watts. So, for the battery drain associated with a 5 mW transmitter, the effective range for a UWB system is more like that of a 5 Watt transmitter. This equates to a lessened load on batteries, and longer battery life for a fixed size battery.

Put another way, whereas a tactical radio might have 90 minutes of talk time on a typical battery, if UWB technology were used instead, talk time, *ceteris paribus*, would approach tens up to hundreds of hours for the same battery charge. Alternately, for a given talk-time, the size of the phone and the cost of the tactical radio could be greatly reduced. Whereas battery technology is mature, and greatly increased battery capacity is not feasible with known battery chemistries, UWB modulation could provide the equivalent effect of a disruptive technological breakthrough in battery technology for implementing a new generation of body-worn, battery-powered communications gear.

1.2   UWB DESCRIPTION AND VENDORS

1.2.1   Description

The history of UWB dates to the earliest days of radio, and to even before radio was called radio, back to when radio was first called *wireless*.[10] Recent advances in digital processing have made it possible to re-think the fundamental trades long used for implementing radios, allowing improvements over the trades when analog circuits were the sole means by which to fashion communication system building blocks. With a fresh re-thinking of communication system implementations arising with UWB technology, it becomes possible to gain significant advantages over previous communication systems implementations, while simultaneously reducing implementation complexity, physical volume, and power consumption.

How is this re-thinking of implementation details, long established by practice, possible? It is possible because UWB communication is simply traditional radio or wireless technology with a different choice of ranked importance of the variables than what has traditionally been chosen. Specifically, UWB communication systems trade pulse shortness, thereby gaining high peak powers, in exchange for two other variables:

1.)   *Bandwidth* (the needs of which are increased in UWB due to the short duration of the pulses), and

2.)   *Signal to noise ratios* of individual pulses (which are decreased in UWB, thereby requiring correlation to combine coherent pulse energies coherently, thereby gaining an advantage over noise powers that only can combine non-coherently, being uncorrelated.)

Some refer to UWB communication as *impulse radio*. Others see it as simply being traditional radar modulation used for communication purposes. Both viewpoints are technically correct.

With a re-thinking of the rules that have governed radio design for so long, UWB technology enables new communication systems to be created with higher performance levels than have ever before been possible.

---

[10] Terrence W. Barrett, *History of UltraWideBand (UWB) Radar & Communications: Pioneers and Innovators*, Progress in Electromagnetics Symposium 2000 (PIERS2000), Cambridge, MA, July 2003. See: http://www.ntia.doc.gov/osmhome/uwbtestplan/barret_history_(piersw-figs).pdf (Retrieved 19 August 2003.)

1.2.2   <u>UWB Vendor Survey</u>

To understand the range of possibilities inherent with UWB technology, it is worthwhile to explore first the major commercial applications being investigated today, prior to tabulating current UWB work by vendor. These possibilities include:

Fade-free Tactical Radios: High-bandwidth tactical radios, providing video and voice, with position-aware features for tracking position in real-time while also providing communication data links.

Localizers: Devices for enabling the real-time tracking location of high-value items to within centimeters on assembly area floors independent of GPS signals that typically are unable to penetrate buildings.

Cable-HDTV Upgrades: Both wireless and wired possibilities exist for UWB technology. For example, shown under Pulse~link is a wired UWB application, enabling the emergence of HDTV overlaid onto existing cable-TV service while eliminating the obsolescence of existing cable-TV equipment.

Perimeter radars: Protection of high-value items through detecting intrusion of people or small robotic instruments.

Long Battery-life Portable Wireless Apparatus: The efficiency of UWB transmitters can increase the effectiveness of existing battery technology.

UWB Chipsets: Fabless semiconductor designers are at work, designing the core chipsets needed by all UWB product designers.

Clearly, this set of possibilities will grow as UWB technology matures, and more possibilities are envisioned. Today, UWB technology is still in its infancy.

In addition to the vendors tabulated in Table 1.2.1, considerable original work has also been done at national laboratories and universities around the United States (e.g., LLNL (Lawrence Livermore National Laboratory), University of Southern California, Clemson University, etc.). Original work has also been done at foreign facilities, especially in the Soviet Union/Russian Federation, Singapore, and China. Despite the international development of UWB technology, this report primarily focuses on just work and products that have either been performed or sold within the United States within the private sector. This is because UWB is a dual-use communication technology, and only companies with a significant presence in the United States will likely support the creation of future Spaceport and Range communication networks. Only these companies have been tabulated in Table 1.2.1, which lists the major UWB vendors active in the UWB market within the US over the last few years.

Table 1.2.1 UWB Vendors & Technical Approaches (Summer 2003)

Company	Location	Modulation	Products	Status
Aether Wire & Location, Inc. www.aether wire.com/	Sunnyvale, CA	Pairs of positive and negative TH-PPM pulses called *doublets*	Pager-sized localizers & comm. devices.	Founded 1991, conducted two years of self-funded R&D. First round of private financing in 1993. Developed first chips in 1.2 micron double-poly CMOS with Orbit Semiconductor in July 1994. $1.8M DARPA grant in 1998. First UWB patent in 1998. Puts spectral nulls where needed (e.g., GPS bands) without filtering through adjusting the spacing between positive and negative pulses. Typically, Aether Wire UWB systems are non-coherent at RF frequencies.
Alereon, Inc.	Austin, TX	Multi-Band TH-PPM	UWB chips	Founded by former Time-Domain Corporation executives; company was first announced August 25, 2003. (Not to be confused with AMD's 1999 K7 microprocessor chip that was also named Alereon.) Has taken over development of the 802.15.3a chipset from Time Domain Corporation known as PulsON 300 or P300.
Cellonics http://www.cell onics.com/inde x.htm	Singapore	Direct PPM UWB through non-linear upconversion without any VCO or mixer required.	Pulse-based Neural Nets. Non-linear UWB processing cells are based on biological analogies.	Founded Jan 1, 2000. First round VC financing May 2000. Holds US patent awarded on first basis (no prior art.) Very inexpensive UWB transmitters available now (Aug 2003). Simplified carrier-rate decoding modules are also available.

Company	Location	Modulation	Products	Status
Discrete Time Communication	San Diego, CA	Unknown	Fabless CMOS ICs for UWB	First ICs were planned Q1 2004; Staccato Communications acquired Discrete Time Communications
Fantasma	San Diego, CA	Unknown	None	$11.6M first-round VC funding in January 2000. Unable to raise 2nd round VC funding; assets purchased by Pulse~link in May 2001. Some senior staff joined Discrete Time Communications.
Farr Research, Inc. www.farr-research.com	Albuquerque, NM	Products to support all times of UWB modulation	UWB antennas, passive UWB components, time-domain antenna ranges, TEM sensors, Electronic Warfare (EW) antennas for using Marx Generators (400 kV pulses)	Numerous UWB antennas and antenna-related products spanning 150 MHz to 20 GHz are available. A catalog of products is available. Major products include collapsible and solid Impulse Radiating Antennas (IRAs), and calibrated TEM (Transverse Electro-Magnetic) wave sensors. UWB antennas for fixed, parachuted, space, and terrestrial uses are available. Much research is conducted with the U.S. Army Space & Missile Defense Command and with Phillips Laboratory, Kirtland AFB, NM.
Furaxa	Orinda, CA	Various UWB modulations through generating UWB pulses with programmable amplitude, position, & duration	Pulser Sampler ICs based on Libove Gates	Libove Gate architecture provides 4+ GHz repetition rate vs. only 250 MHz in earlier Gilbert Cell or Schottky Bridge + step recovery diode (SRD) pulser sampler architectures. Programmable UWB feature permits changing modulation details to meet evolving or new FCC rule changes 'on the fly.'

Company	Location	Modulation	Products	Status
General Atomics	San Diego, CA	Multi-band OFDM (Spectral Keying^{tm})	480 Mb/s IAW IEEE 802.15.3a	Founded in 1955. Photonics division is assigned responsibility for developing UWB. Teamed with Philips, to use Philips QuBIC semiconductor processes based on their own Spectral Keying^{tm} technology.
General Electric http://www.crd.ge.com/	US, India, China	Various UWB modulations	Delay-hopped transmitted reference UWB comm.	GE Global Research has 2,000 researchers working in three research labs in the US, India, and China.
Harris Corporation (Government Communications Systems)	Palm Bay, FL	Bi-Phase Pulse	UWB defense products	Teamed with XSI, and uses their chipsets in producing UWB products for defense applications.
Intel (Intel Architecture Labs, (IAL))	Hillsboro, OR	2-PAM with high PRF	LAN/PAN applications. Presumably will ultimately support IEEE 802.15.3a.	Focused on MAC, and data transport issues at present, placing less emphasis on PHY layer than seen with many other UWB companies. Likely to depend on just acquiring a start-up to acquire a complete PHY layer capability once UWB standards mature and stabilize. Potential candidates would be Staccato or perhaps XSI. See: www.intel.com/technology/itj/q22001/articles/art_4.htm
I-tech	Slovenia	Unknown	Tx/Rx	Products available now.
Motorola (Semiconductor Products Sector)	Austin, TX	Bi-Phase Pulse	UWB consumer electronics & computing market products (e.g., WPANs IAW IEEE 802.15.3a) *(planned)*	Teamed with XtremeSpectrum on March 10, 2003 to produce UWB consumer products using XSI's UWB Trinity^{tm} chipsets.

Company	Location	Modulation	Products	Status
Multispectral Solutions, Inc. (MSSI)	Germantown, MD	FDM-TDMA UWB	Defense (Military) in comm., radar, geo-positioning areas (various). Tactical (1-2 km) as well as strategic (>100 km) UWB systems.	Founded 1988. Has developed UWB handheld transceivers, UWB radar altimeter, UWB sources, and UWB intrusion detectors. Has won over 60 UWB contract awards. Wireless LPI LPD intercoms/headsets (WICS) transitioned to production in July 2003. Typically, MSSI's UWB systems are non-coherent at RF frequencies.
ParthusCeva, Inc.	San Jose, CA. Dublin, Ireland (Parthus) for RF technology + Santa Clara, CA (Ceva) for DSP cores	DS-spread pulse signaling from 3.85 to 7.7 GHz with bi-orthogonal M-ary symbols constructed using ternary Golay-Hadamard sequences, in combination with Reed-Solomon and convolutional error-control coding	55 to 980 Mb/s (proposed)	Parthus Technologies PLC merged with Ceva, Inc. on September 26, 2002 upon a shareholder vote to merge the two operations. (Ceva, Inc. was formerly a subsidiary of US firm DSP Group, Inc.) ParthusCeva, Inc. ownership: DSP Group (fabless semiconductor company) owns 50.1%; Parthus owns 49.9%.
Royal Philips Electronics	Amsterdam, the Netherlands	Multi-band OFDM	Up to 480 Mb/s UWB chipsets IAW IEEE 802.15.3a	Based on Philips' QUBiC semiconductor processes (e.g., QuBIC3 is a low-cost 0.5 micron 70 GHz f_{max} silicon BiCMOS process). Using license of General Atomics' spectral keying technology (i.e., multi-band OFDM UWB).
Pulse~link	San Diego, CA	MPEG DVD transport over UWB over wireline	UWB at 400 Mb/s up to 10 meters, 7 Mb/s up to 100 meters, both over wireline.	Founded June 2000 in Panama City, FL. Moved to San Diego, CA with purchase of Fantasma's assets. First to demonstrate UWB over wired media. Intends to be *the* HDTV CATV upgrade provider by 2005. Developing a very large UWB patent portfolio.

Company	Location	Modulation	Products	Status
Pulsicom	Israel	Unknown	Unknown	Intel Forum, Oct 11, 2001
Samsung Electronics Co., Ltd.	Korea	Multi-band OFDM	480 Mb/s UWB consumer, mobile, & computing PAN products IAW IEEE 802.15.3a	Partnered with Staccato to use Staccato's ICs. (Press release 12 August 2003)
Skycross, Inc. www.skycross.com/	Melbourne, FL	Products to support all times of UWB modulation	UWB Antennas	Has several designs for meeting UWB needs. Early products achieved operating bandwidths over just 3.1 to 6.0 GHz, as a sub-set of the current 3.1 to 10.6 GHz Part 15 regulations. Skycross holds a significant meanderline antenna patent portfolio (much of which is from BAE) that can achieve extended UWB-sized bandwidths while keeping physical antenna volumes small.
Staccato Communications	San Diego, CA	Multi-band OFDM	CMOS UWB ICs.	A fabless producer of UWB ICs. Formerly was Discrete Time Communications.
STMicroelectronics	Multiple locations; multiple countries	Position & polarity modulation with convolutional or turbo error-control coding, occupying 3 to 7 GHz	62.5 to 500 Mbps UWB ICs (proposed)	40,000 employees in 27 countries. Company was formed in June 1987 as a result of a merger between SGS Microelettronica of Italy and Thomson Semiconducteurs of France. Invested $977.9M (15.4% of revenues) in R&D in 2001.
Texas Instruments	Dallas, TX	Multi-band 128-tone OFDM using 528-MHz bands and QPSK for the tone modulation	55 to 480 Mbps (proposed)	Patents currently exist for this PHY approach; some licensing workarounds will be needed if this proposed modulation is selected as the 802.15.3a standard.

Company	Location	Modulation	Products	Status
Time-Domain Corporation (TDC)	Huntsville, AL	TH-PPM (positive pulses, only)	UWB Chipsets, Radarvision^{tm} Through-wall radar, Eval Kits.	Shipping Evaluation Kits, Radarvision^{tm} units. Typically, TDC UWB systems are coherent at RF frequencies, providing performance advantages.
Taiyo Yuden (TRDA)	Tokyo, Japan; USA: Chicago, San Jose, San Marcos, Dallas, & Raleigh	Bi-phase Pulse	UWB modules (planned)	TRDA is the USA-based research and development arm of Taiyo Yuden. Teamed with XSI to produce UWB modules.
WisAir www.wisair.com/	Tel-Aviv, Israel	Multi-band variable rate PHY for IEEE 802.15.3a	20 to 125 Mb/s UBLink^{tm} chipsets and antennas. Evaluation toolkit (available June 1, 2003	UBLink^{tm} chips support 1-15 sub-bands selectable out of 30. WisAir successfully demonstrated transport of multiple HDTV streams using UWB on June 20, 2003 in Tokyo, Japan.
XtremeSpectrum Incorporated (XSI) www.xtremespectrum.com/	Vienna, VA; bay area, CA	Bi-Phase Pulse	UWB Chipsets (Trinity^{tm})	Founded 1998, and produced many of the early UWB chipsets used for defense applications. Trinity chipset launched June 2002. Evaluation Kit & UWB chips were due out July 2003, but slipped. XSI is teamed with Harris Corporation for defense applications. XSI teamed with TRDA/Taiyo Yuden on January 9, 2003 to produce UWB modules. XSI teamed with Motorola March 10, 2003 to produce UWB products.

1.3   BASIC UWB THEORY

The following introduces UWB theory starting with the simplest monocycle representation that incorporates all the fundamentals necessary for understanding basic UWB communication principles. Then, additional levels of detail are added as necessary for building on these principles for introducing more esoteric UWB concepts. The general approach chosen is to start with the representation of a monocycle seen at the output of a receive antenna, and to base all the correlation calculations on this most commonly used representation of a received monocycle. As UWB theory is expanded, different correlation templates are derived.

The preliminary introduction, in turn, is followed by a discussion of higher levels of complexity in the monocycle waveform itself, through examining the monocycle waveform (1) as it is produced as a Gaussian current pulse, (2) as it is transmitted from the transmitter antenna, (3) as it is received through the receive antenna, and (4) as it becomes a current pulse that is processed by the receiver. Understanding this time-domain complexity leads to the recognition of a *theory of relativity* as applied to monocycles. Namely, an observed monocycle changes its time-domain shape depending upon where the particular UWB monocycle is observed in a UWB system.[11]

Comparisons of monocycles with other solitary waves (solitons, wavelets) are also introduced where necessary for comparing and contrasting the spectral characteristics of these solitary waves with monocycles.

Likewise a new technology application is developed for detecting UWB transmissions without requiring any *a priori* knowledge of the parameters of the UWB monocycles to be detected. This new technology application is based on wavelets, and provides a new, powerful method for detecting otherwise difficult-to-detect, illicit, or otherwise covert, UWB transmitters, such as used for electronic bugging purposes.

1.3.1   Simplified Monocycle Introduction

Traditional wireless radio transmissions have utilized sinusoidal waveforms since the 1920's for a variety of reasons. Perhaps the most compelling reason has been that sinusoidal waveforms are very amenable to mathematical modeling. Another reason is that, because of this ease of analysis, sinusoidal waveforms also make the analytical task easier for reducing occupied communication system bandwidths to near the minimum Nyquist-limit bandwidths required for such transmissions, thereby increasing spectral occupancy efficiency and permitting more transmitters to occupy the airwaves without causing one another harmful interference.

---

[11] Since this report is primarily focused on communications systems, UWB theory is not developed in this report beyond that which is required for understanding UWB communication principles. Further theoretical investigations, into UWB ground penetration and through-wall radar monocycle principles, remain topics for future research.

What happens, though, if these traditional trades are re-ranked by a new set of priorities? Unlike traditional wireless signals, UWB transmissions do not attempt to reduce their occupied bandwidth. Instead, UWB transmissions work to increase their occupied bandwidths to values much greater than the required minimums. This trade is intentionally made for improving communication link performance while permitting lower average transmitter power levels to be used. UWB communication therefore permits, and even requires, a re-selection of the trades that have been traditionally chosen, while overcoming many historical performance limits.

The basic UWB waveform is an approximation of a Gaussian pulse. Specifically, UWB radio systems use Time-Hopping (TH) nanosecond (or shorter) duration Pulse Position Modulated (PPM) pulses known as monocycles to propagate signals over physical distances instead of the sinusoidal carriers used by most radio systems. A typical time-domain representation of a UWB monocycle waveform pulse is shown in Figure 1.3.1 as a received monocycle, and is simply a first-order approximation to an ideal Gaussian pulse waveform. This particular representation of a waveform assumes no channel distortions, and represents an observation of an idealized UWB monocycle, *p(t)*, observed approximately six inches from an appropriately ultra-wideband transmitting antenna, at the output of a second ultra-wideband receiving antenna, from which it is observed as a received current pulse.[12] There is undershooting on both the leading and trailing edge of this waveform. This particular UWB waveform is based in large part on a typical empirically-selected normalizing monocycle width value, $\tau_n = 0.4472$, selected for a best fit to a particular waveform monocycle by Ramirez-Mireles and Scholtz.[13]

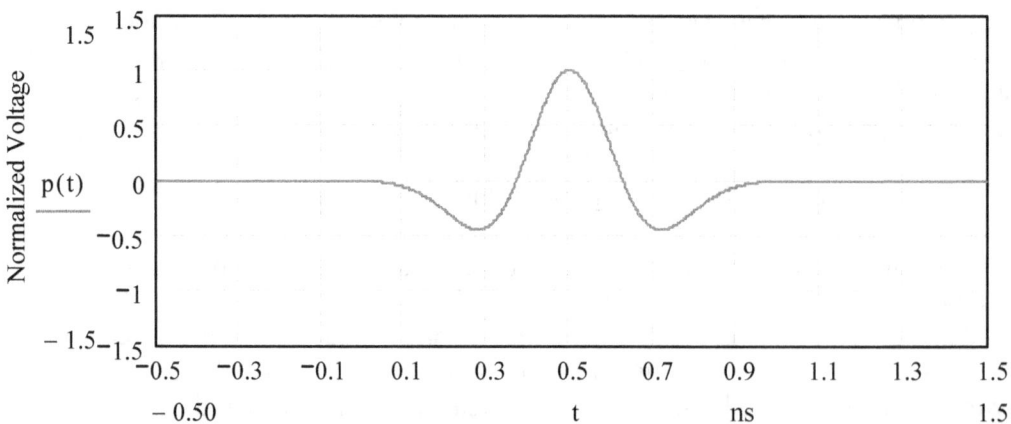

Figure 1.3.1 A Single TH-PPM Monocycle Pulse

---

[12] There is an additional level of subtle complexity that will be introduced later in this paper, in which it will be shown that the monocycle waveform discussed here is *not* preserved throughout the UWB system. Rather, the shape, and hence the spectral characteristics, of monocycles differ depending upon where the monocycles are observed.

[13] Fernando Ramirez-Mireles and Robert A. Scholtz, "System Performance Analysis of Impulse Radio Modulation", IEEE Proceedings RAWCON Conference, August 1998. http://ultra.usc.edu/New_Site/publications.html (University of Southern California Ultra Lab).

Mathematically, this monocycle time function shape can be expressed as the following:[14]

$$ p(t) := \left[1 - 4 \cdot \pi \cdot \left(\frac{t - \tau d}{\tau n} \right)^2 \right] \exp\left[-2\pi \cdot \left(\frac{t - \tau d}{\tau n} \right)^2 \right] $$

Monocycles provide practical and implementable waveforms that greatly improve data rate versus power consumption trades compared to traditional sinusoidal radio waveforms. For short (fixed) communication distances, monocycles enable communication at very high data rates at very low power consumption. Likewise, monocycles support the determination of relative location information among a network of receivers and transmitters. Monocycle waveforms also can be used to enable precise inspection and geo-location functionality for Ground Penetration radar systems.

Monocycles first arose in classified, non-ground penetration, radar applications. One of the first uses was for target discrimination in cluttered environments (e.g., searching for aircraft over ocean expanses, or searching for vehicles embedded within foliage). There were also other early uses for monocycles for achieving aircraft identification through taking time domain responses of radar reflections.[15] The significant fundamental theories for monocycles were derived almost entirely within the context of military radar systems.

In the simplest, earliest radar systems, *individual* monocycles or pulses always had to exceed a threshold for detection; in current radar systems and in even the simplest UWB systems, the sum totals of collections of pulses must always exceed a noise threshold, although individual pulses often are often well below noise thresholds.[16] Monocycles hence support achieving processing gain, similar to that achieved in spread spectrum communication systems. This is true regardless of whether monocycles are used within radar systems or within UWB communication systems. This characteristic also often allows the successful use of lower power levels than would otherwise be possible.

Unlike in fixed radar installations, UWB communication applications are ill suited for use in radio links having significant Doppler shifts. The reason is that determining time references becomes very difficult for deciding bit decisions 'on the fly' between ZEROs and ONEs in a continuous running UWB communication link, with closing or separating physical distances changing at high rates. (This will be shown later, while discussing the correlation detection process for demodulating digital data.) This deficiency could be addressed, through the incorporation of more elaborate decoding techniques, but at the expense of worsened complexity in the UWB receiver circuitry. This would negate a key advantage of UWB communication systems in the typical communication application,

---

[14] Fernando Ramirez-Mireles and Robert A. Scholtz, "System Performance Analysis of Impulse Radio Modulation*", IEEE Proceedings RAWCON Conference, August 1998. http://ultra.usc.edu/New_Site/publications.html (University of Southern California Ultra Lab).

[15] C. E. Baum and E. G. Farr, Impulse Radiating Antennas, H. L. Bertoni (eds.), pp. 139-147 in *Ultra-Wideband, Short-Pulse Electromagnetics*, New York, Plenum, 1993.

[16] Mischa Schwartz, *Information Transmission Modulation and Noise, A Unified Approach to Communication Systems,* McGraw-Hill, New York, NY, 1959, p. 409.

namely, simplicity. In a high Doppler environment, the normal simplicity advantage of UWB radios would be largely lost due to increases in decoding complexity.

1.3.2   Detection of UWB Monocycles

Because correlation is most often used for demodulation of monocycles, there are mathematical properties that monocycle waveforms must absolutely meet in order for convergence and proper detector correlation processor operation to occur in a UWB receiver.[17] Specifically, it is necessary that the function of the UWB monocycle integrated over all time (i.e., its area) be finite, and additionally equal to zero, in order for the correlation integral to converge. Namely, the monocycle waveform requirement is that:

$$A1 := \int_{-\infty}^{\infty} p(t)\, dt$$

must be numerically equal to, and must evaluate to, zero, which it does for the selected *p(t)* function given previously.

Although coherent detection processing is most commonly used in UWB receivers to provide the highest levels of performance, non-coherent processing (at RF) is sometimes also used to lower the recurring costs of UWB hardware for applications where cost matters more than performance.[18] The advantage of coherent detection processing is that an individual UWB monocycle modeled as p(t) can be coherently detected, even when many signals comprise a broadband noise floor that buries the desired monocycle signal in a cacophony of interference. Non-coherent processing, on the other hand, requires higher signal levels, and/or a lessened interference environment for the successful detection of non-coherent monocycles.

---

[17] Robert A. Scholtz, P. Vijay Kumar, and Carlos J. Corrada-Bravo, "Signal Design for Ultra-wideband Radio", Sequences and Their Applications (SETA '01), Bergen, Norway, May 13-17, 2001. (Work sponsored by Office of Naval Research under grant N00014-96-1-1192 (subcontract of the Univ. of Puerto Rico), and by the National Science Foundation under grant ANI-9730556.)

[18] Not all vendors chose coherent processing in designing their UWB receivers. Among the major vendors, only Time-Domain has always used coherent RF processing. Others, such as Aether Wire and Multispectral Solutions, typically have not used coherent RF processing. The new 802.15.3a standard being developed will likely require coherent RF processing. Coherent processing provides the highest functionality and is the most extensible. Non-coherent processing achieves the lowest cost, at the penalty of meeting only lower performance, with severely limited functionality and extensibility. See: Paul Withington, *"Ultra-Wide Band Radio, A New Frontier"*, Singapore IDA UWB *Programme* Framework, 25 February 2003.

Continuing with the highest performance, coherent processing method, consider one normalized signal correlation function, *γp(t)*, given by Ramirez-Mireles and Scholtz that can be used to detect the previously defined p(t):[19]

$$\gamma p(t) := \left[1 - 4 \cdot \pi \cdot \left(\frac{t}{\tau n} \right)^2 + \frac{4 \cdot \pi^2}{3} \left(\frac{t}{\tau n} \right)^4 \right] \exp\left[-\pi \cdot \left(\frac{t}{\tau n} \right)^2 \right]$$

Graphically, this normalized UWB signal correlation template function resembles the monocycle it detects, although there are slight differences in the shape of the template from the monocycle.

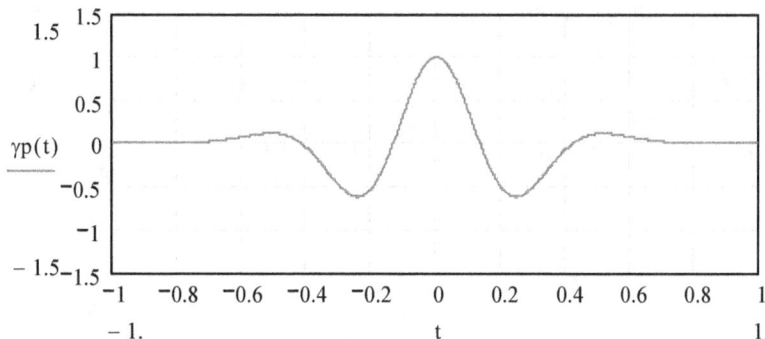

Figure 1.3.2 A TH-PPM Monocycle Correlation Template Function

Similar to the requirement that exists on the UWB monocycle waveform itself for convergence of the detection process, it is also necessary that the integrated value over all time of the UWB signal correlation template function (i.e., its area) likewise be both finite and equal to zero. Namely, it is necessary that:

$$A2 := \int_{-\infty}^{\infty} \gamma p(t) \, dt$$

when evaluated numerically, be equal to zero, which is true for this UWB signal correlation template function.

1.3.3 UWB Correlation Gain Constant

In order to calculate the appropriate scaling factor for the correlation process, it is necessary to determine the minimum value of the correlation function. This can be done easily through solving:

$$\frac{d}{dt} \gamma p(t) = 0$$

to find the inflection points in the correlation function, i.e., through solving:

---

[19] Fernando Ramirez-Mireles and Robert A. Scholtz, "System Performance Analysis of Impulse Radio Modulation", IEEE Proceedings RAWCON Conference, August 1998.

$$\frac{d}{dt}\left[\left[1 - 4\cdot\pi\cdot\left(\frac{t}{\tau n}\right)^2 + \frac{4\cdot\pi^2}{3}\left(\frac{t}{\tau n}\right)^4\right]\exp\left[-\pi\cdot\left(\frac{t}{\tau n}\right)^2\right]\right] = 0$$

Solving this equation, it is found that there are five numerical solutions:

$$\begin{pmatrix} 0 \\ 1.1397661323729298411\cdot\tau n \\ -1.1397661323729298411\cdot\tau n \\ .54081659961081661355\cdot\tau n \\ -.54081659961081661355\cdot\tau n \end{pmatrix}$$

Since $\tau n = 0.4472$ for this running example, the minimum of interest is easily found to be at $\tau min = 0.2419$ ns. The evaluated value of $\gamma p(\tau min) = -0.6183$. With this correlation function minimum, the normalized gain correction value, β, of the received PPM equally correlated signals is next calculated: [20]

$$\beta := \frac{1 + \gamma p(\tau min)}{2}$$

For the value $\gamma p(\tau min) = -0.6183$, this means that the beta-factor, β, = 0.1909.

What is this beta-factor? It is simply the nominal gain of the detector correlator in the receiver. It is used through defining a normalized correlation between the UWB monocycle and the UWB correlation factor, including the beta-factor, as:

$$\text{Det}(t) := \frac{1}{\beta}\cdot\int_{-\infty}^{\infty} p(\tau)\cdot\gamma p(t - \tau)\, d\tau$$

Examining this equation, the beta-factor is seen to be the normalizing, or scaling, correlator gain needed to set the peak value of the output of the correlator so that its detected output is equal to the original received monocycle amplitude. Verifying this numerically, consider the peak values of both the detector correlator output and the original monocycle amplitude at the center of the detected waveform that is physically offset from the broadband UWB antenna:

$\text{Det}(\tau d) = 1.00208$ $p(\tau d) = 1.00000$ As expected the two values are the same, at least to three digits.

This can also be seen through examining a plot of the correlator detector's output, scaled by the beta-factor, when plotted against the original UWB monocycle, $p(t)$. (Note that the sidelobes of the correlator detector's output are shaped slightly different than the original UWB monocycle. This side-lobe difference must be considered in practice,

---

[20] Fernando Ramirez-Mireles and Robert A. Scholtz, "System Performance Analysis of Impulse Radio Modulation", IEEE Proceedings RAWCON Conference, August 1998.
http://ultra.usc.edu/New_Site/publications.html (University of Southern California Ultra Lab).

since in a multi-path environment, these "sidelobes" can easily become confused in some UWB receiver implementations with correlations resulting from other arriving, though weaker, multi-path-traveled monocycles.)

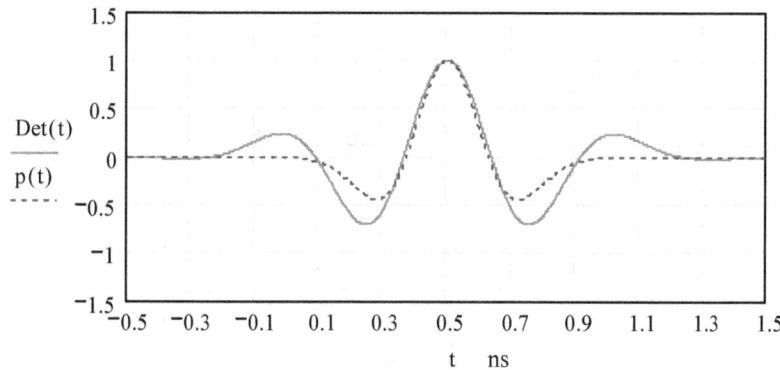

Scaled, Detected, Received Correlator Output, Det(t), vs. UWB Monocycle, p(t)

Figure 1.3.3 Comparing Scaled, Detected, Received Output vs. Received UWB Monocycle

1.3.4   Monocycle Pulsetrain Analysis

The UWB monocycle pulse equation can also be written simpler to aid in analyzing a pulse train of monocycles. As shown previously, the normal waveform used to represent a single UWB monocycle is expressed as:

$$p(t) := \left[1 - 4 \cdot \pi \cdot \left(\frac{t - \tau d}{\tau n} \right)^2 \right] \exp\left[-2\pi \cdot \left(\frac{t - \tau d}{\tau n} \right)^2 \right]$$

For convenience, we now simplify the subsequent pulse train constants, following Huang's simplification method.[21] Let T = 3, and

$$a := \frac{2 \cdot \pi}{\tau n^2}$$

---

[21] Po T. Huang, NASA KSC, private correspondence and discussions.

The monocycle function, p(t), can then be re-written as g(t), where g(t), with the simplification becomes:

$$g(t) := \left[1 - 2a \cdot (t - T)^2 \right] \cdot \exp\left[-a \cdot (t - T)^2 \right]$$

Plotting this function:

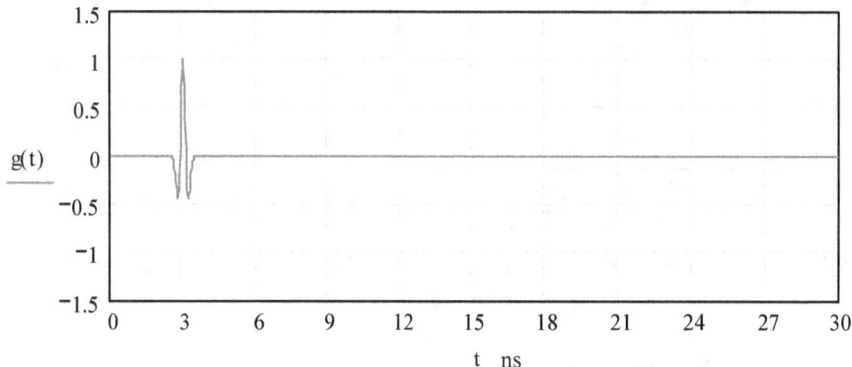

Figure 1.3.4 A Single TH-PPM Monocycle Pulse of a Pulse Train

This plot shows a single monocycle, centered on the first period centered at 3 ns.

1.3.5   UWB Spectral Analysis

Although UWB monocycles are often Pulse Position Modulated, a significant amount of insight into the fundamental spectral characteristics of UWB transmissions can be derived from assuming a regularly spaced pulse train of unmodulated UWB monocycles. Although there is more regularity on the spectral nulls with this simplifying assumption, the basic spectral envelope is still largely retained.

A pulse train, h(t), of UWB monocycles spaced regularly in time, can be modeled as:[22]

$$h(t) := \sum_{n=1}^{m} \left[\left[1 - 2 \cdot a \cdot (t - n \cdot T)^2 \right] \cdot \exp\left[-a \cdot (t - n \cdot T)^2 \right] \right]$$

---

[22] Po T. Huang, NASA KSC, private correspondence and discussions.

In the time domain, this can be plotted, giving:

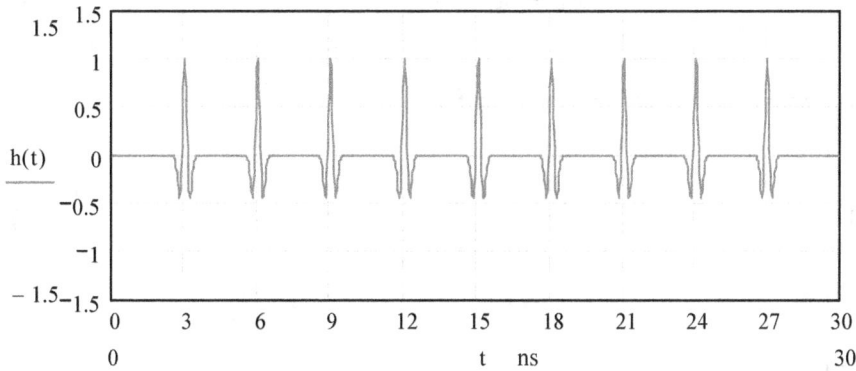

Figure 1.3.5-1 A TH-PPM Monocycle Pulse Train

The spectrum of the amplitude of this pulse train can be modeled as a Fourier Transform of the time function, h(t):[23]

$$F(\omega) := \int_{1.5}^{28.5} h(t) \cdot e^{-2 \cdot \pi \cdot 1i \cdot \omega \cdot t} \, dt$$

For best accuracy, adaptive integration instead of Romberg integration should be used for evaluating this integral.

The pulse train's power spectrum can be modeled in dBW-scaled terms (i.e., in dB) (cf. Parseval's Theorem for a detailed rationale for why there should be a 2π term in the denominator) as:

$$P(\omega) := 10 \cdot \log\left(\frac{|F(\omega) \cdot F(-1i \cdot \omega)|}{2 \cdot \pi} \right)$$

---

[23] H. Joseph Weaver, Theory of Discrete and Continuous Fourier Analysis, Wiley Inter-science, New York, NY, 1989.

Plotting this equation, the spectral occupancy over ultra wide bandwidths becomes obvious, with significant roll-off evident over frequencies below the peak of the power spectral density. There is an even faster rate of spectral roll-off, of 20 dB to 40 dB per octave, observed over higher frequencies.

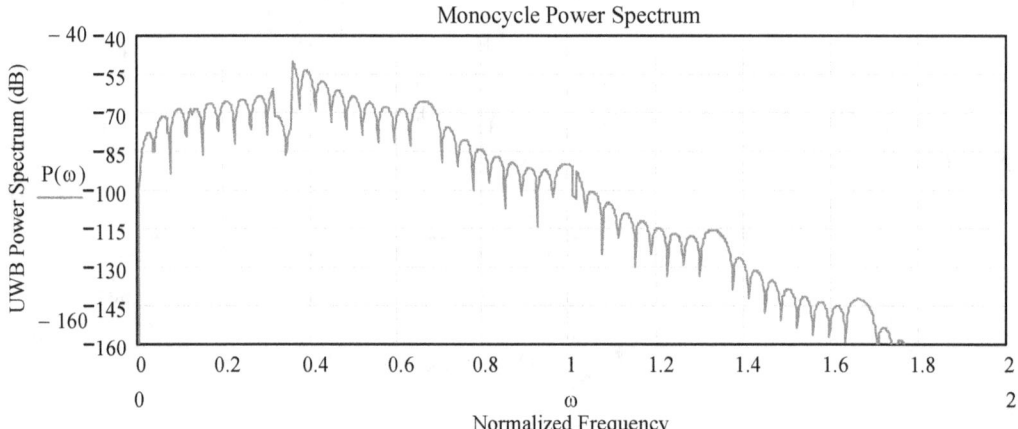

Figure 1.3.5-2 A TH-PPM Monocycle Pulse Train Power Spectrum

Of course, this modeled spectrum represents the all-ZERO case, which contains no TH-modulation. If unmodulated and modulated monocycles were both present in the pulse train, the actual spectrum would shift, partially filling the multiple spectral nulls seen in this plot.

1.3.6   <u>Finding a UWB Correlation Template through Cross Correlation</u>

The first correlation template function used previously is but one such template function that can be used successfully to detect UWB monocycles. It is also possible to generate a successful correlation template function through first taking a derivative of a monocycle time function and then applying a cross correlation technique. (Think of this as a double-step process for creating a correlation template.) The details of applying this technique can be demonstrated as follows, starting first with the same time function representing a monocycle in the time domain as before:

$$p(t) := \left[1 - 4 \cdot \pi \cdot \left(\frac{t - \tau d}{\tau n} \right)^2 \right] \exp \left[-2\pi \cdot \left(\frac{t - \tau d}{\tau n} \right)^2 \right]$$

For one correlation template model, it is possible to set the pulse correlator waveform, $\omega cor(t)$, equal to the derivative of this monocycle time function; that is:[24]

$$\omega cor(t) := \frac{d}{dt}p(t)$$

The monocycle cross correlation function, $R\omega(t)$, is then the cross correlation between the received monocycle and the pulse correlator waveform:

$$R\omega(\tau) := -\int_{-\infty}^{\infty} p(t + \tau) \cdot \omega cor(t)\, dt$$

The slope at time zero of this new function can be easily computed:

$$m(\tau) := \frac{d}{d\tau}R\omega(\tau) \qquad\qquad m := m(0) \qquad\qquad y(t) := m \cdot t$$

Plotting this cross correlation function:

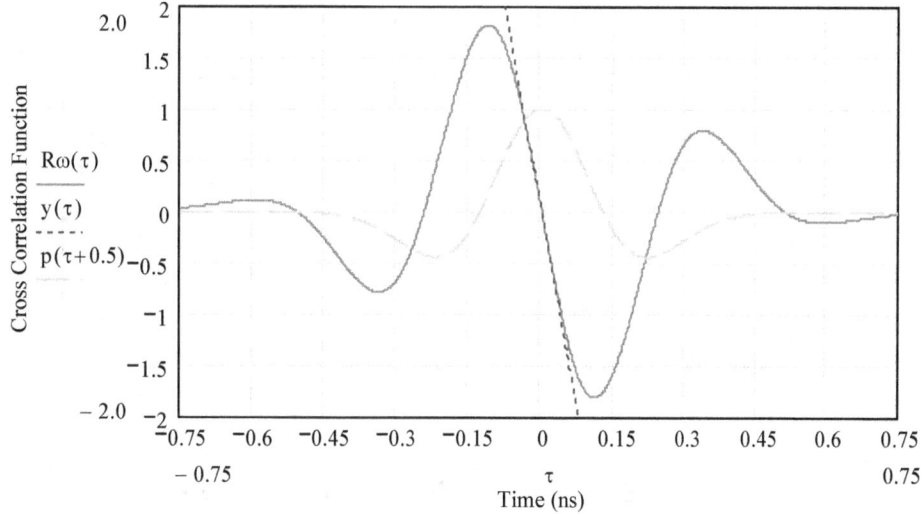

Figure 1.3.6 A Cross-Correlation TH-PPM Monocycle Correlation Template

This template function derived from a template function derived from a derivative of the monocycle waveform will be subsequently used and shown later to work as one more acceptable correlation template function for the detection of UWB monocycles.

---

[24] Moe Z. Win and Robert A. Scholtz, "Ultra-wide Bandwidth Time-Hopping Spread-Spectrum Impulse Radio for Wireless Multiple-Access Communications", IEEE Trans. Comm. Vol. 48, No. 4, April 2000.

1.3.7   Comparison of UWB Monocycles vs. Exponentially-shaped Cos Pulses

Many spectral analyses assume a pulse train of cosine pulses to simulate UWB transmissions. Unfortunately, spectral emissions predicted this way differ significantly over some frequencies than when UWB monocycles are assumed. Looking first at a single cosine pulse, and using Huang's simplification method as before, assume a cosine pulse of the form:[25]

$$\cos(b \cdot t) \cdot e^{-a \cdot t^2}$$

Setting up the constants the same as before:

$$\tau n = 0.44720 \qquad a := \frac{2 \cdot \pi}{\tau n^2} \qquad b := 0.5 \cdot a \qquad m := 9 \qquad T := 3$$

And setting up the pulse   train as before:

$$g1(t) := \sum_{n=1}^{m} \left[\cos\left[b \cdot (t - n \cdot T) \right] \cdot \exp\left[-a \cdot (t - n \cdot T)^2 \right] \right]$$

This function can be plotted as:

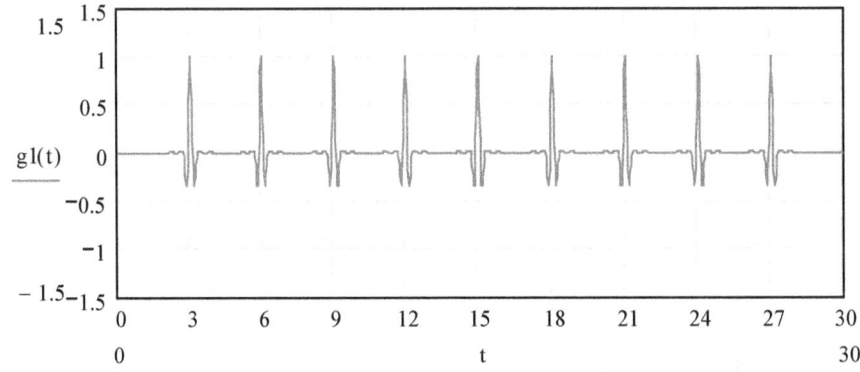

Figure 1.3.7-1 Exponentially-shaped Cosine Pulse Train

---

[25] Po T. Huang, NASA KSC, private correspondence and discussions.

At first glance, this greatly resembles the monocycle pulse train previously analyzed. But, consider whether this resemblance is more than just an illusion through additionally examining the power spectrum of this waveform. The spectrum of the amplitude of the pulse train, F2(ω), can be modeled as a Fourier Transform of the time function g1(t):[26]

$$F2(\omega) := \int_{1.5}^{28.5} g1(t) \cdot e^{-2 \cdot \pi \cdot 1i \cdot \omega \cdot t}\, dt$$

The power spectrum can be modeled in dBW-scaled terms (in dB) as:

$$P2(\omega) := 10 \cdot \log\left(\frac{\left| F2(\omega) \cdot F2(-1i \cdot \omega) \right|}{2 \cdot \pi} \right)$$

Graphically comparing this power spectrum against the earlier one done for a pulse train of monocycles results in the following:

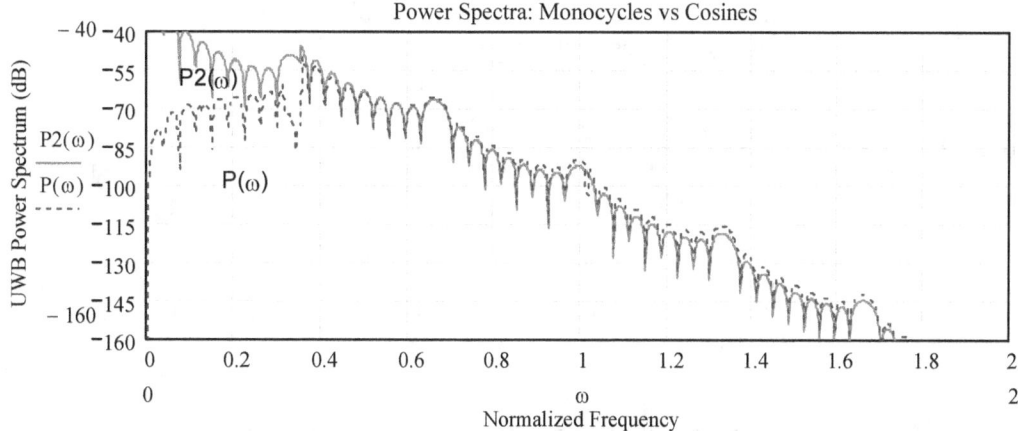

Figure 1.3.7-2 Cosine Pulse Train Spectrum vs. Monocycle Power Spectrum

This result is illuminating, for though the Gaussian monocycle and exponentially shaped cosine pulse are very similar in the time domain, their power spectrums differ significantly over the lower normalized frequencies. Modeling an FCC Part 15 UWB transmission with exponentially shaped cosine pulses could lead to inferring erroneously that the interference potential against GPS signals from FCC Part 15 UWB transmissions occupying 3.1 GHz to 10.6 GHz would be much higher at the 1.575 GHz L1 GPS band than would actually be the case. Especially when predicting the interference susceptibility of lower frequency, narrowband emissions to UWB transmissions, and when accurately modeling spectral emissions from summing numerous UWB

---

[26] H. Joseph Weaver, Theory of Discrete and Continuous Fourier Analysis, Wiley Inter-science, New York, NY, 1989.

transmissions simultaneously, real monocycle waveforms should be used instead of sinusoidal pulses for modeling UWB emissions over all frequency ranges.

In short, a Gaussian pulse (i.e., a monocycle) shape matters significantly when making accurate spectral emission predictions, especially over frequencies below the peak of the power spectrum. (The preservation of information that occurs with Gaussian pulses when taking derivatives, as occurs when a monocycle passes through the transmit antenna and again when passing through the receive antenna, is also not preserved with exponentially shaped cosine pulses; this is an even stronger reason not to use sinusoidal pulses for modeling UWB transmissions.)

1.3.8   <u>Second-order Introduction to Monocycles</u>

Why is the commonly used monocycle model what it is? The most commonly used monocycle model is simply a Gaussian pulse modified to add negative sidelobes on the leading and trailing edges, unlike a true Gaussian pulse that does not 'ring' at all.[27] This modification simply incorporates what occurs, and is seen, with actual hardware.

A finite series approximation to an ideal Gaussian pulse, pb(t), can be derived through truncating an infinite Gaussian series such as given by Zverev as follows:[28]

$$pb(t) := \left[1 + 2 \cdot \left(\frac{t - \tau d}{\frac{\tau n}{\sqrt{2 \cdot \pi}}} \right)^2 + \frac{2^2}{2!} \cdot \left(\frac{t - \tau d}{\frac{\tau n}{\sqrt{2 \cdot \pi}}} \right)^4 + \frac{2^3}{3!} \cdot \left(\frac{t - \tau d}{\frac{\tau n}{\sqrt{2 \cdot \pi}}} \right)^6 + \frac{2^4}{4!} \cdot \left(\frac{t - \tau d}{\frac{\tau n}{\sqrt{2 \cdot \pi}}} \right)^8 + \frac{2^5}{5!} \cdot \left(\frac{t - \tau d}{\frac{\tau n}{\sqrt{2 \cdot \pi}}} \right)^{10} \right] \cdot \exp\left[-2 \cdot \pi \cdot \left(\frac{t - \tau d}{\frac{\tau n}{\sqrt{2 \cdot \pi}}} \right)^2 \right]$$

This is an idealized Gaussian pulse with no undershoot on either the leading or trailing edge. When plotted, this lack of undershoot can be seen:

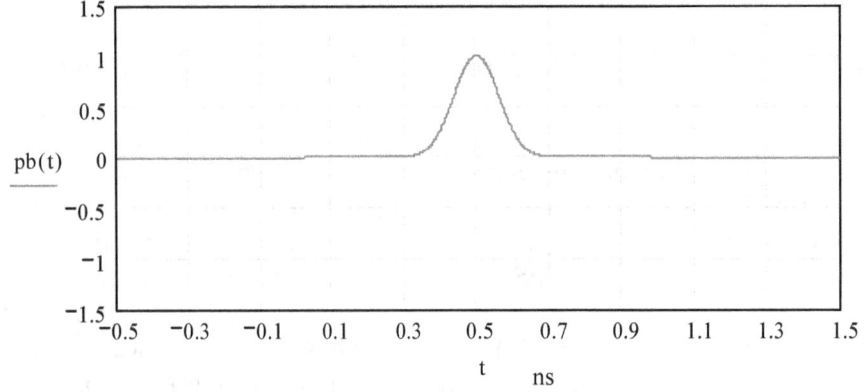

Figure 1.3.8-1 Gaussian Pulse Exhibits No Undershoot

---

[27] Paul Withington, formerly of Time-Domain Corporation, Huntsville, AL, private discussions. Mr. Withington left Time Domain in May 2003 while this analysis was being developed.
[28] Anatol I. Zverev, *Handbook of FILTER SYNTHESIS,* John Wiley & Sons, New York, 1967, pp. 70-71.

It is possible to add a leading and trailing edge undershoot through manipulating this equation, subtracting the truncated Gaussian series from "2", which normalizes the peak amplitude at unity, while also adding a scaling factor in the exponential shaping factor.[29] The result of this equation manipulation is:

$$pa(t) := \left[2 - \left[1 + 2 \cdot \left(\frac{t - \tau d}{\frac{\tau n}{\sqrt{2 \cdot \pi}}} \right)^2 + \frac{2^2}{2!} \cdot \left(\frac{t - \tau d}{\frac{\tau n}{\sqrt{2 \cdot \pi}}} \right)^4 + \frac{2^3}{3!} \cdot \left(\frac{t - \tau d}{\frac{\tau n}{\sqrt{2 \cdot \pi}}} \right)^6 + \frac{2^4}{4!} \cdot \left(\frac{t - \tau d}{\frac{\tau n}{\sqrt{2 \cdot \pi}}} \right)^8 + \frac{2^5}{5!} \cdot \left(\frac{t - \tau d}{\frac{\tau n}{\sqrt{2 \cdot \pi}}} \right)^{10} \right] \right] \cdot \exp\left[-\left(\frac{t - \tau d}{\frac{\tau n}{\sqrt{2 \cdot \pi}}} \right)^2 \right]$$

This model has severe undershoot. However, through truncating this series approximation to but a first order approximation, as in pa1(t) which follows, the undershoot can be lessened considerably.

$$pa1(t) := \left[2 - \left[1 + 2 \cdot \left(\frac{t - \tau d}{\frac{\tau n}{\sqrt{2 \cdot \pi}}} \right)^2 \right] \right] \cdot \exp\left[-\left(\frac{t - \tau d}{\frac{\tau n}{\sqrt{2 \cdot \pi}}} \right)^2 \right]$$

Simplifying further, this equation is easily seen to be just the standard equation that is often published in the literature for a UWB monocycle, and which has been discussed already at length in this paper:

$$p(t) := \left[1 - 4 \cdot \pi \cdot \left(\frac{t - \tau d}{\tau n} \right)^2 \right] \exp\left[-2\pi \cdot \left(\frac{t - \tau d}{\tau n} \right)^2 \right]$$

The standard UWB monocycle model is therefore just a modified Gaussian pulse, modified slightly through the inclusion of a 2π scaling factor in the exponential weighting factor to add a slightly negative undershoot to the waveform model to match what is seen in practice with real hardware, which occurs because of limitations to bandwidth.

---

[29] Anatol I. Zverev, *Handbook of FILTER SYNTHESIS*, John Wiley & Sons, New York, 1967, pp. 70-71.

Comparing these different steps in building a model of a UWB monocycle waveform, the following plots show the relationships more clearly:

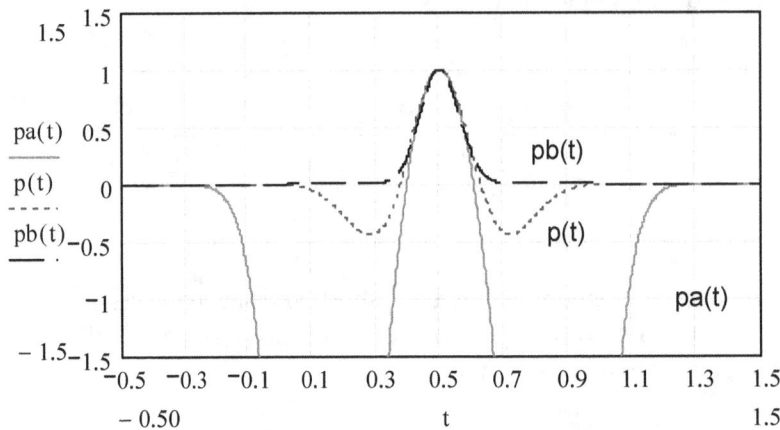

Figure 1.3.8-2 Gaussian Pulse Approximations Show Undershoot

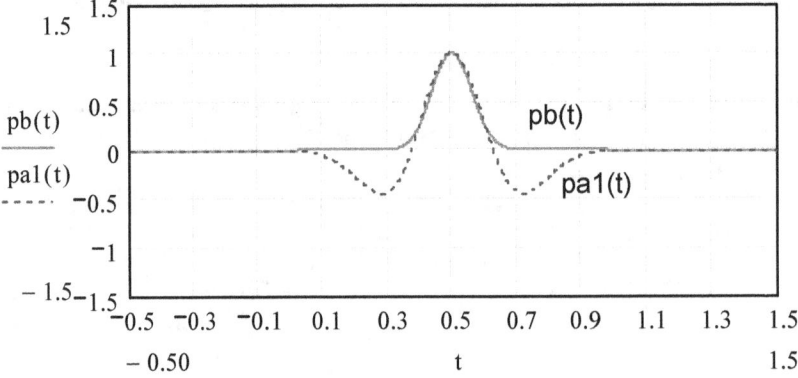

Figure 1.3.8-3 First Order Gaussian Pulse Approximation has Minimal Undershoot

(Although this discussion explains why the UWB monocycle model is what it is, there is still considerably more subtle theory that must be developed before a more complete UWB monocycle picture becomes evident. Specifically, still remaining is the effect of antennas on a non-continuous signal impulse. For explaining this effect, additional monocycle theory will be developed shortly, in considerably more detail, in a detailed third-order (third-level) discussion of monocycles, to be introduced shortly after the theoretical details of optimal matched receivers and the theoretical comparisons of monocycles with other solitary waves are developed.)

1.3.9   Optimal Matched Receiver for UWB Monocycles

The matched receiver that represents the optimal receiver for a single bit of a binary-modulated impulse radio signal in Additive White Gaussian Noise (AWGN) is a correlation receiver, using an appropriately matched correlation template signal, v(t).[30] The simplest matched correlation template signal is v(t) = p(t) – p(t-δ), assuming δ is appropriately chosen so as to minimize inter-symbol interference (ISI) between ONEs and ZEROs.[31,32] That is, for:

$$p(t) := \left[1 - 4\cdot\pi\cdot\left(\frac{t - \tau d}{\tau n}\right)^2 \right] \exp\left[-2\pi\cdot\left(\frac{t - \tau d}{\tau n}\right)^2 \right]$$

v(t) becomes

$$v(t) := \left[1 - 4\cdot\pi\cdot\left(\frac{t - \tau d}{\tau n}\right)^2 \right]\cdot\exp\left[-2\cdot\pi\cdot\left(\frac{t - \tau d}{\tau n}\right)^2 \right] - \left[1 - 4\cdot\pi\cdot\left(\frac{t - \delta - \tau d}{\tau n}\right)^2 \right]\cdot\exp\left[-2\cdot\pi\cdot\left(\frac{t - \delta - \tau d}{\tau n}\right)^2 \right]$$

The optimized value of δ to minimize ISI for the specific p(t) model listed above has been found previously to be δ = 0.156.[33]

The simplest matched correlator template signal, v(t) is shown graphically in the following:

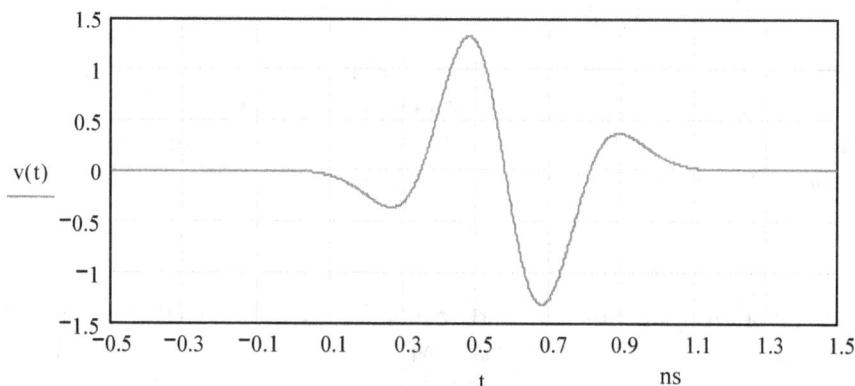

Figure 1.3.9-1 An Optimal Matched Correlation Template

---

[30] Robert A. Scholtz and Moe Z. Win, "Impulse Radio", Invited Paper, IEEE PIMRC'97, Helsinki, Finland.
[31] Ibid.
[32] Moe Z. Win and Robert A. Scholtz, "Ultra-wide Bandwidth Time-Hopping Spread-Spectrum Impulse Radio for Wireless Multiple-Access Communications", IEEE Trans. Comm. Vol. 48, No. 4, April 2000.
[33] Ibid.

This template signal will be used extensively in later analyses in this paper.

Another commonly used template signal found in the literature is the appropriately scaled and time-shifted derivative of the monocycle itself, as alluded to earlier in this paper.[34] To understand why this works as well as it does, consider the following scaled and time-shifted function:

$$v1(t) := 0.12 \cdot \left(\frac{d}{dt} p(t - 0.08) \right)$$

This template signal, v1(t), when plotted against the optimal template signal, v(t), shows that a striking resemblance exists:

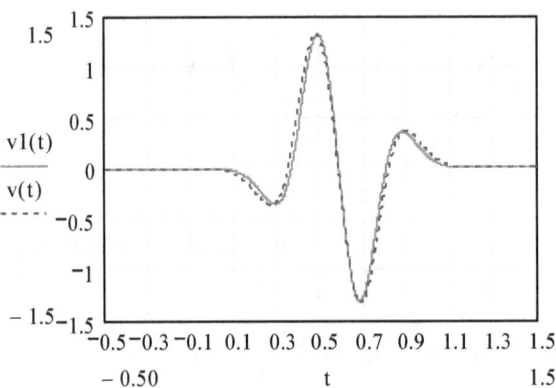

Figure 1.3.9-2 Approximating an Optimal Matched Correlation Template

Recall that the monocycle cross correlation function $R\omega(\tau)$ is the cross correlation between the received monocycle and the pulse correlator waveform. Appropriately scaled and time-shifted, it is likewise another loose approximation to the same correlator function approximations already shown. Specifically, consider:

$$Rw(t) := 0.74 \, R\omega(t - 0.58)$$

Plotting this new function against the two earlier functions v(t) and v1(t) shows a striking resemblance among all three. Because of this, Rw(t) is another template signal that can be used for detecting monocycles. The choice of which template signal to use, among the many shown in this paper, will often depend on implementation advantages of one of these signals specifically in an actual application, instead of on any great advantage of performance. They all will work with similar performance in most applications.

---

[34] Moe Z. Win and Robert A. Scholtz, "Ultra-wide Bandwidth Time-Hopping Spread-Spectrum Impulse Radio for Wireless Multiple-Access Communications", IEEE Trans. Comm. Vol. 48, No. 4, April 2000.

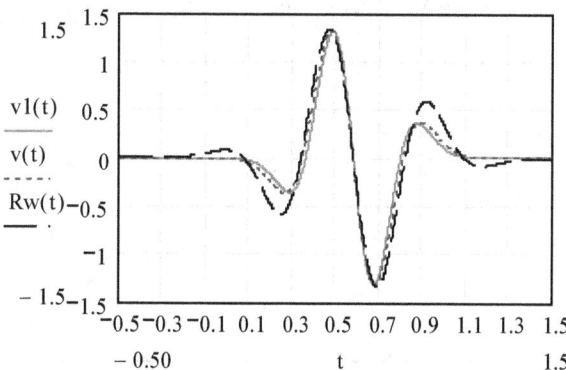

Figure 1.3.9-2 Approximations for an Optimal Matched Correlation Template

1.3.10   Pulse Position Modulating UWB Monocycles

With the theory developed thus far, it becomes possible to model monocycles as well as the Time Hopped Pulse Position Modulation (TH PPM) of the same monocycles. Recall that Time Hopping/Hopped (TH) PPM is one of the two major modulations applied to UWB communication systems.[35]

In TH PPM UWB modulation, multiple monocycles are used to transmit each bit in each symbol. The monocycles that comprise each bit, which in turn comprise each bit symbol, must individually be detected, and soft bit decisions must be made, prior to making a final bit symbol decision. Consider a UWB TH PPM modulation scheme whereby a monocycle occurring at exactly the repetition period is defined to be a "ZERO" and a delayed monocycle, delayed by δ (the value of which is chosen to minimize ISI, that is, Inter-Symbol Interference), is defined to be a "ONE". Assume a period of 3, i.e., T = 3, for the following example. Furthermore, assume that somehow, perhaps with a delay-locked loop (DLL), time synchronism is achieved. A ZERO would be transmitted

---

[35] The other major modulation used is Bi-Phase monocycle modulation. There are often practical advantages for TH PPM over Bi-Phase modulation, especially for keeping spurious spectral emissions down in RF amplifiers used to amplify UWB transmissions.

The issue is that, during the phase cross over in Bi-Phase modulation, non-linear (more efficient) RF amplifiers often become unstable, emitting spurious emissions. The fix is relatively easy, but is not easily implemented in production for large numbers of RF amplifiers, requiring the addition of Select-At-Test (SAT) values of capacitance across transmitting semiconductor junctions to allow specific frequency harmonics to have a return path to ground at the RF amplifier's output. The inclusion of a low pass filter with a Pi-configuration in place of a T-configuration also accomplishes this, but this often restricts desired UWB emissions, causing a slight mismatch and reduction in operating bandwidth in order to squelch the spurious emission.

It was hoped that this phenomenon could be investigated on this project, but no evaluation hardware was available from XtremeSpectrum, Inc. (XSI), who uses Bi-Phase modulation exclusively in their UWB designs. Because of this, only the TH PPM modulation was analyzed extensively on this project and in this paper.

through repetitively transmitting the following, for a modulation, m, with m = 0 for transmitting a ZERO:

$$p(t) := \left[1 - 4 \cdot \pi \cdot \left(\frac{t - m \cdot \delta}{\tau n} \right)^2 \right] \exp \left[-2\pi \cdot \left(\frac{t - m \cdot \delta}{\tau n} \right)^2 \right]$$

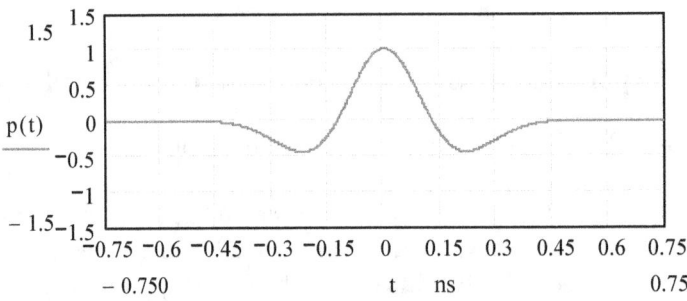

Figure 1.3.10-1 Transmitting a ZERO

Likewise, for transmitting a ONE, m = 1, and the following, transmitted repeatedly, would comprise a ONE bit:

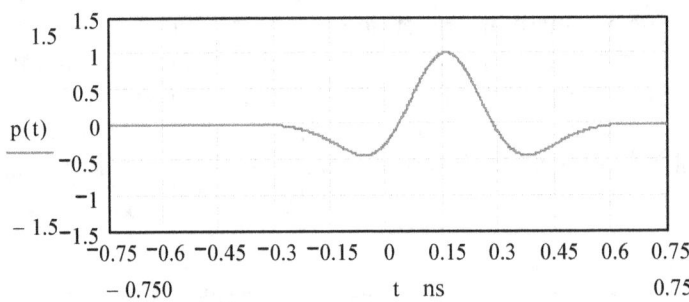

Figure 1.3.10-2 Transmitting a ONE

Depending on when a repeated monocycle arrives, it is possible to determine whether a ONE or a ZERO monocycle were transmitted. This exact determination is accomplished through a correlation of the monocycle waveform against the correlation template signal.

To see the detection process in terms of the mathematics, consider the transmission and subsequent reception of a single monocycle, modulated as either a ZERO or as a ONE, as shown previously. A signal, *s*, based on the correlation between the p(t) monocycle and the correlation template is compared with zero. If s is greater than zero, then a ZERO is *decisioned*. Likewise, if *s* is less than zero, then a ONE is *decisioned*. This can be demonstrated as follows. Set m = 0, thereby intending to transmit a ZERO, and assume p(t) as shown previously. The signal, s, is therefore defined as:[36]

$$s := \int_0^T p(t) \cdot v(t)\, dt$$

Numerically evaluating this integral, $s = 0.01272$. Since s is greater than zero, this is decisioned as a ZERO.

Instead, set m = 1, thereby intending to transmit a ONE and assume p(t) as shown previously. The signal, s, is therefore still defined as:

$$s := \int_0^T p(t) \cdot v(t)\, dt$$

Numerically evaluating this integral, $s = -.06719$. Since s is less than zero, this is decisioned as a ONE.

The significance of v(t), defined as the difference between the monocycle and a delayed version of itself, is important. If correlators were 'free', it would theoretically be possible to perform two correlations, i.e., perform a correlation between p(t) as received versus p(t) as ideal, and perform another correlation between p(t) as received and p(t-δ). Then, depending on which correlation were greater, it would be possible to assess whether a ZERO or ONE had been transmitted.

But, correlators implemented in hardware or in firmware are not 'free'. There is always an implicit cost associated with implementing correlators. Because of this, it is more economical to form a v(t) correlation template from the difference between two states, thereby reducing the correlation hardware required by a factor of 2. This is even more significant when implemented multiple times in parallel in real hardware, for it likewise becomes possible to reduce the number of correlators required per each parallel detection bin in an actual receiver implementation to just one instead of two. A 50% reduction in correlation hardware that spans multiple bins is well worth the complexity of developing a correlation template waveform that requires a one-time subtraction of one waveform from another.

---

[36] Predrag Spasojevic and Arashk Mahjoubi Amine, "Ultra wide band Time-Hopping Modulation for Multiple users", Rutgers University, http://www.caip.rutgers.edu/~arashk/UWB.pdf (http://www.eden.rutgers.edu/~arashk/), retrieved April 2003.

It is also possible to use one of the other correlation templates, derived earlier, to perform the decisioning. Recall the v1(t) template, defined earlier. It is possible to compute a signal, *s*, using this template, too:

$$s := \int_0^T p(t) \cdot v1(t) \, dt$$

For m = 1, *s* can be numerically evaluated, as *s* = -0.05084, which is decisioned as a ONE. Likewise, for m = 0, *s* can likewise be numerically evaluated, as *s* = 0.01308, which is decisioned as a ZERO. As this shows, it matters very little which template signal is used for performing the correlation, provided that a valid template that approximates a bit-duration correlator is used.

Now, a single monocycle is all that has been detected. As introduced earlier, a number of such monocycles actually comprise each bit.[37][38] What is the effect of including a large number of monocycles (say 10,000) for transmitting repeatedly a single bit? The effect is simply to provide a more robust decisioning process, as the few soft decisions that are in error, perhaps from processing during a burst of noise that somehow manages to approximate a slipped monocycle, affect the hard decision process at the conclusion of all the soft monocycle decisions only slightly. Whereas a single ONE monocycle is (for m = 1):

$$p(t) := \left[1 - 4 \cdot \pi \cdot \left(\frac{t - m \cdot \delta}{\tau n} \right)^2 \right] \exp \left[-2\pi \cdot \left(\frac{t - m \cdot \delta}{\tau n} \right)^2 \right]$$

a pulse train of P monocycles (for, say P = 9), that likewise comprise part of a ONE bit, can be modeled as:

$$h(t) := \sum_{n=1}^{P} \left[\left[1 - 2 \cdot a \cdot (t - m \cdot \delta - n \cdot T)^2 \right] \cdot \exp \left[-a \cdot (t - m \cdot \delta - n \cdot T)^2 \right] \right]$$

---

[37] Predrag Spasojevic and Arashk Mahjoubi Amine, "Ultra wide band Time-Hopping Modulation for Multiple users", Rutgers University, http://www.caip.rutgers.edu/~arashk/UWB.pdf http://www.eden.rutgers.edu/~arashk/ , retrieved April 2003.
[38] Moe Z. Win and Robert A. Scholtz, "Ultra-wide Bandwidth Time-Hopping Spread-Spectrum Impulse Radio for Wireless Multiple-Access Communications", IEEE Trans. Comm. Vol. 48, No. 4, April 2000.

which, when plotted, becomes:

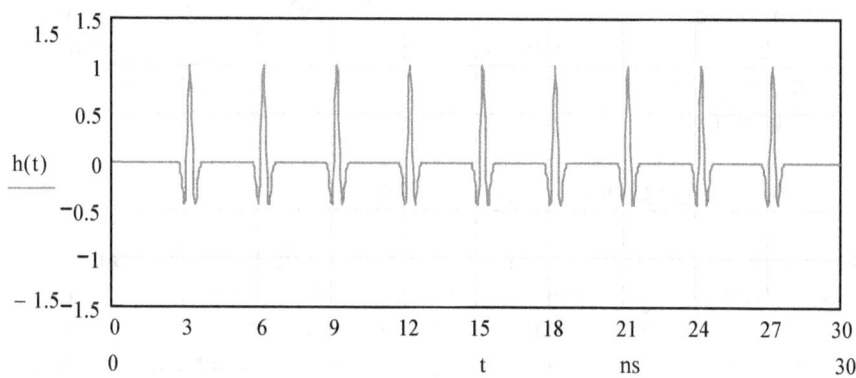

Figure 1.3.10-3 Transmitting a ONE over Multiple Monocycles

This is therefore a ONE bit-symbol, for the case where only 9 monocycles comprise a bit. Note especially the slight offset, thereby constituting a ONE modulation, in the peaks of the monocycles versus the repetition periods every 3 ns.

In a noiseless environment, a good approximation to the detection process for P consecutive monocycles is:

$$s := P \cdot \int_0^T p(t) \cdot v(t) \, dt$$

Evaluating this numerically, the signal, s, is found to be $s = -0.60472$. Being less than zero, this is decisioned as a ONE. Clearly, though, this is a more robust evaluation than the integral resulting from the correlation involving only a single ONE monocycle.

Likewise, as for the case of only a single monocycle, it is possible to use a different correlation template signal in this detection process, also. Consider the use of the $\gamma p(t)$ correlation template signal, defined earlier. Using this template signal, s is found as:

$$s := P \cdot \int_0^T p(t) \cdot \gamma p(t) \, dt$$

This, evaluated numerically, evaluates to $s = -0.1509$. Of course, this is decisioned as a ONE bit. This still works, albeit with a probable less robust detection in the presence of noise, due to the computation of a smaller value (indicating a less robust template signal in many environments, no doubt.)

Similarly, the evaluation of ZERO monocycles can be performed, for m = 0, in the same equations.

In the absence of noise, all of the correlation templates would undoubtedly work well. With noise, or with other UWB monocycle pulse trains, however, the optimization of the correlation template could provide considerable benefit for increasing the robustness of the communication link.

1.3.11 <u>Maximizing Disorder for a Fixed Standard Deviation</u>

As introduced earlier, the presence of noise can greatly diminish the robustness of a non-optimal communication link. Likewise, the modeling of noise can greatly influence the apparent robustness of a communication link, especially if the noise that is modeled is not representative of the actual noise properties that will be encountered. The need for finding the 'noisiest noise' is therefore a real one, if the modeling is to reflect the actual performance likely to ensue with particular correlator architectures.

Finding the probability density function that maximizes the uncertainty or measure of disorder (i.e., the entropy) for a one-dimensional probability distribution function (PDF), P(x), with a fixed standard deviation, σ, is important for selecting the noisiest noise model for simulating random variables. (The following derivation largely follows the same method used by Claude Shannon in his original analysis of entropy for communication systems.[39]) The entropy of a Random Variable (RV) x is given by:

$$H(x) = \int P(x) \cdot \log(P(x), 2) \, dx$$

where log (P(x), 2) indicates the base-2 logarithm of P(x).

The standard deviation for a Random Variable x is given by:

$$\sigma^2 = \int P(x) \cdot x^2 \, dx$$

Now, the integral of the PDF must equate to unity, if the PDF accurately states the locations where the Random Variable x can be. That is:

$$1 = \int P(x) \, dx$$

[39] C. E. Shannon, "A Mathematical Theory of Communication", Reprinted with corrections from The Bell System Technical Journal, Vol. 27, pp. 379-423, 623-656, July, October, 1948, based on the BSTJ version as opposed to the University of Illinois Press 1949 reprint version, http://cm.bell-labs.com/cm/ms/what/shannonday/paper.html, retrieved 3 June 2003, p. 36.

Maximizing the entropy, H(x), by the calculus of variations, specifically through using a Legendre Transformation, requires maximizing:[40]

$$\int \left(-P(x) \cdot \log(P(x), 2) + \lambda \cdot P(x) \cdot x^2 + \zeta \cdot P(x) \right) dx$$

But, $0 <= P(x) <= 1$ since $P(x)$ is a PDF and max $(P(x)) = 1$. This means that the following condition must hold:

$$-1 - \log(P(x), 2) + \lambda \cdot x^2 + \zeta = 0$$

Adjusting the constants λ and ζ to satisfy the previously listed constraints and properties, $P(x)$ is found to be:

$$P(x) = \frac{1}{\sigma \cdot \sqrt{2 \cdot \pi}} \cdot \exp\left[-\left(\frac{x^2}{2\sigma^2} \right) \right]$$

for the case where the probability distribution function is centered on zero. For the case where a shifted PDF, centered around a mean, μ, is desired, shifted to the right along the x-axis, the property of maximized entropy for this PDF still holds, but the functional form must be modified slightly; substituting $(x-\mu)$ for x in $P(x)$, $P(x)$ becomes:

$$P(x) = \frac{1}{\sigma \cdot \sqrt{2 \cdot \pi}} \cdot \exp\left[\left(\frac{-1}{2} \right) \cdot \left(\frac{x - \mu}{\sigma} \right)^2 \right]$$

This is the same model that will be used next to simulate noise.[41]

As an interesting aside, a monocycle itself is a Gaussian pulse that closely resembles this same function. Because of this, and by the same reasoning as above, the entropy of a monocycle waveform is very nearly optimized. The impact of this is that a monocycle is an excellent waveform shape for implementing radio links using the lowest power levels that are theoretically possible, since maximizing pulse entropy maximizes the monocycle's own distinction from the noise in which it arrives. For this reason, UWB is likely to be one of the dominant waveforms for future radio links in which the lowest possible transmitted power for a given data rate is desired. This presumes, of course, that co-interference and non-interference issues with traditional narrowband signals are resolved.

---

[40] I. M. Gelfand and S. V. Fomin, Calculus of Variations, Dover, Mineola, NY, 1991, pp. 71-75.
[41] Howard W. Sams & Co., Inc, *Reference Data for Radio Engineers*, 6th Edition, 2nd Printing, 1977, New York, NY, p. 42-4.

1.3.12 Using Normal Distributions to Model and Explore Noise and Path Loss

There are two key advantages to using a Normal Distribution for modeling natural processes. First, as derived previously, Normal Distributions model natural processes well since they maximize entropy, which natural processes tend to maximize. Second, Normal Distributions are easy to use, since it is possible to define probability density functions (PDFs) through defining just two parameters, namely the mean, μ, and the standard deviation, σ, or equivalently the entropy, H, in place of the standard deviation. Knowing just two values, therefore, anyone can reproduce an entire PDF, with all of its properties.

For example, arbitrarily set the two key parameters to define completely a specific Normal Distribution as $\sigma = 0.5$, and $\mu = 1.0$ Now, the entropy, which is the measure of disorder, is given by:[42]

$$H := \frac{\log\left(\sigma \cdot \sqrt{2 \cdot \pi \cdot e}\right)}{\log(2)}$$

which, when evaluated, gives an entropy H = 1.04710 bits for the chosen value of σ.

1.3.13 Detecting Monocycles in a Noisy, Variable Attenuation Channel

This same technique can be applied to the problem of detecting monocycles in a noisy, variable attenuation channel. For this, assume that a Standard Deviation for Gaussian Noise of 0.25 is used; that is, $\sigma = 0.25$. Likewise, assume gain constants of Vg = 0.3 and Mg = 0.62 (both chosen through iteratively trying various values in the following equations until the resulting noise matched experimental expectations of real noise.)

Next, randomize the mean of the noise to implement a highly variable channel attenuation, even within a monocycle, when simulated by double sideband (DSB) modulating a monocycle waveform, p(t), with noise. (This variability just approximates the spectral correlation nulls that occur over a few MHz of bandwidth at various points within the ultra wide bandwidths occupied by the UWB monocycle.)

For the following simulation, a random number generator function was used to generate a mean: $\mu = $ rnd (1) * 0.6 where rnd (1) returns a random value ranging from 0 to 1, and hence a mean, μ, ranging from 0.0 to 0.6. (For the exact value returned for one sample run, a mean of 0.43456 was obtained for use in the following equations.)

---

[42] C. E. Shannon, "A Mathematical Theory of Communication", Reprinted with corrections from The Bell System Technical Journal, Vol. 27, pp. 379-423, 623-656, July, October, 1948, based on the BSTJ version as opposed to the University of Illinois Press 1949 reprint version, http://cm.bell-labs.com/cm/ms/what/shannonday/paper.html, retrieved 3 June 2003, p. 37.

Now, define a noisy monocycle signal with amplitude p(t), lightly DSB modulated by scaled AWGN to simulate a fast-changing channel, in a scaled additive noise channel (also AWGN) with a mean of μ, and a standard deviation of σ:[43]

$$\text{Signal}(t) := p(t) \cdot \frac{Mg}{\sigma \cdot \sqrt{2 \cdot \pi}} \cdot \exp\left[\left(\frac{-1}{2}\right) \cdot \left(\frac{\text{rnd}(0.8 \cdot \sigma) - \mu}{\sigma}\right)^2\right] + \frac{Vg}{\sigma \cdot \sqrt{2 \cdot \pi}} \exp\left[\left(\frac{-1}{2}\right) \cdot \left(\frac{\text{rnd}(\sigma) - \mu}{\sigma}\right)^2\right]$$

DSB modulation simply translates the simulated noise spectrum onto the monocycle "carrier" for investigating (verifying) the subsequent correlation-based bit-decision operations. Unlike classic DSB modulators, neither the "carrier" nor the data are sinusoidal here. Still the monocycle itself is a periodic waveform, and a useful translation of the simulated noise spectrum is accomplished.

A correlation template v(t), defined the same as before, will be the correlation template chosen. That is, v(t) = p(t) – p(t-δ).

The Signal(t), simulating a monocycle received in a real-world channel, is plotted, as well as a v(t) template signal, as follows:

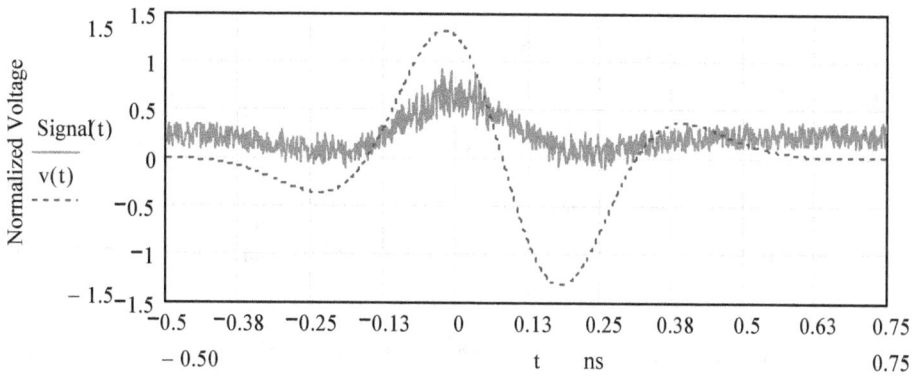

Figure 1.3.13-1 Receiving a ZERO in noise

Shown in the dotted (blue) background is the optimal correlation template signal. Shown in red, overlaying this, is the unmodulated monocycle.

Similar to the correlation technique introduced earlier, this can be decisioned as a ZERO or a ONE through evaluating *s*:

$$s := \int_0^T \text{Signal}(t) \cdot v(t) \, dt$$

---

[43] Howard W. Sams & Co., Inc, *Reference Data for Radio Engineers*, 6th Edition, 2nd Printing, 1977, New York, NY, p. 42-4.

Numerically evaluating this, with adaptive integration, for T = 3 and δ = 0.156, gives *s* = 0.02348, which, being greater than zero, is decisioned as a ZERO.

On the other hand, a noisy ONE monocycle would arrive time-shifted by δ to the right (assuming time synchronization was somehow achieved.) Shifting the same noisy monocycle, to simulate receiving a ONE monocycle in noise, the following results:

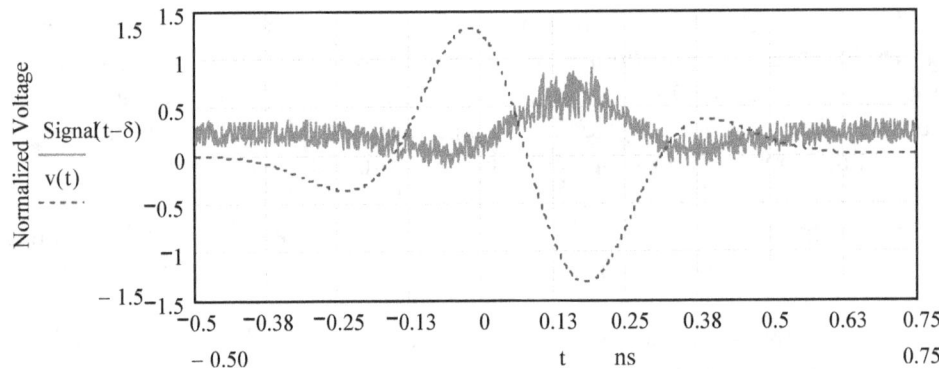

Figure 1.3.13-2 Receiving a ONE in noise

Running the correlation again, to see what would be decisioned:

$$s := \int_0^T \text{Signal}(t - \delta) \cdot v(t)\, dt$$

This signal, *s*, evaluates numerically to *s* = -0.08262, which, being less than zero, is decisioned as a ONE.

Clearly, this correlation technique works to make the soft decisions (ZERO/ONE) for each monocycle that in turn sum to form the decision for the ZERO/ONE bits that in turn comprise the bit symbols that transport data over a TH-PPM UWB communication link.

1.3.14   Monocycle Interference Effects

The UWB signal modeling done thus far has neglected all monocycle distortion effects introduced through the propagation channel, save for the noise and the so-called *flat fade* amplitude effects, which attenuate all frequencies equally. In reality, though, there is distortion that occurs because some frequencies are attenuated more than other frequencies, in clearly what is not a *flat fade*. A differential in path loss versus frequency for the UWB channel introduces dispersion distortion to the perfect monocycle waveform that becomes readily evident on received monocycles.[44] Practical correlation templates must therefore differ significantly from the simple model demonstrated previously if a

---

[44] Ali Taha and Keith M. Chugg, "On Designing the Optimal Template Waveform for UWB Impulse Radio in the Presence of Multipath", 2002 IEEE Conference on Ultra Wideband Systems and Technologies.

proper accounting for dispersion is to be made. They must also be able to distinguish between actual monocycles and other waveforms, as well as between monocycles of multiple UWB transmitters, if UWB communication links are to become practical.

What is the interference potential for a single, exponentially shaped, cosine pulse to be recognized incorrectly as a monocycle?

As before, assume the following constants:

$$a := \frac{2 \cdot \pi}{\tau n^2} \qquad\qquad b := 0.5 \cdot a \qquad \tau n = 0.44720 \qquad T = 3.00000$$

Assume a cosine pulse of the form:

$$pcos(t) := cos(b \cdot t) \cdot e^{-a \cdot t^2}$$

and a monocycle of the same p(t) form.

What will both of these waveforms be decisioned as? For the cosine pulse:

$$s := \int_0^T pcos(t) \cdot \gamma p(t)\, dt$$

Evaluating this numerically, $s = 0.06492$, which being greater than zero, is decisioned as a ZERO.

Likewise, for the monocycle pulse:

$$s := \int_0^T p(t) \cdot \gamma p(t)\, dt$$

Evaluating this numerically, $s = 0.09563$, which being greater than zero, is likewise decisioned as a ZERO.

Graphically, the two ZERO pulses do resemble one another superficially:

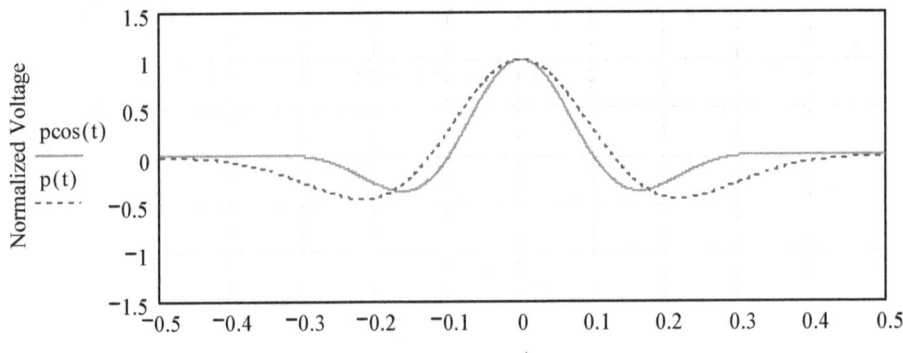

Figure 1.3.14 Comparing a Monocycle with a Cosine Pulse

Clearly, the monocycle is decisioned with a higher correlation, but both are still decisioned as ZEROs.

This points out a potential problem area in future UWB communications at the physical (PHY) layer, once multiple communication links are running simultaneously; namely, the potential exists for confusing individual monocycles transmitted from different UWB transmitters, if their times of arrival at a receiver happens to overlap. Higher-level protocols will likely be required to address this issue, as it is not easily addressable at the PHY layer unless significantly more complexity is used in generating the monocycle waveforms themselves.

1.3.15   Solitary Waves: Solitons vs. Monocycles

Monocycles, used for UWB communications as discussed extensively already, are just one type of solitary wave; other solitary waves exist. Another solitary wave with some of the same properties as a monocycle is a *soliton*. Although all monocycles are solitary waves, and all solitons are solitary waves, solitons and monocycles are not entirely equivalent. Both of these solitary waves, however, do act similarly in distributing energy over ultra wide bandwidths. UWB optical systems, based on 5 fS (i.e., 5 fempto seconds) optical soliton-pulses, have been reported in the literature as occupying ultra wide wavelengths extending from 480 to 835 nm.[45] In the frequency domain, this clearly meets the minimum bandwidth needed to qualify as ultra wideband energy distribution.

Solitary waves, such as monocycles and solitons, can be constructed from a variety of smooth functions. Monocycles, based on Gaussian series expansions (i.e., their smooth function) as discussed at length previously, have specific properties that differ greatly from most solitons in their entropy. UWB monocycle waveforms approximate, as closely

---

[45] Liming Li, Satoru Kusaka, Naoki Karasawa, Ryuji Morita, Hidemi Shigekawa, and Mikio Yamashita, "Amplitude and Phase Characterization of 5.0 fS Optical Pulses Using Spectral Phase Interferometry for Direct Electric Field Reconstruction," Jpn. J. Appl. Physics, Vol. 40, (2001), pp. 684-687, part 2, No. 7A, 1 July 2001.

as can be implemented in practice, the maximum entropy of any communication pulse waveshape. They therefore serve to transmit information with very close to the optimal waveshape for transmitting information over a given distance at the lowest possible transmitted power level. Monocycles are therefore very close to optimal in terms of their power efficiency when compared with traditional communication waveforms that are based on continuous sinusoidal waveshapes.

Solitons are solitary wave solutions to Nonlinear Partial Differential Equations (NPDEs). They are also solitary pulses with many unusual properties. For example, they are waves that act much like particles, preserving their distinctive shapes even when they interact with other solitons. Because of these unusual properties, they are nearly indestructible, traveling for long-distances in water, or in fiber optic cables, for example, while preserving their distinctive shapes. This means that they also exhibit superposition, which in general does not hold for nonlinear systems. Solitons obtain their unusual properties through taking advantage of the dispersion and nonlinear properties of the medium in which they travel. Monocycles, on the other hand, obtain their advantages through management of the entropy of their waveshape, independent of the media in which they travel.

The basic time function that expresses the general form of a soliton solution to a NPDE is given by:[46]

$$u(x,t) = \frac{c}{2} \cdot \text{sech} \left[\frac{\sqrt{c}}{2} \cdot (x - c \cdot t) \right]^2$$

Note that the constant c establishes the amplitude, as well as the velocity of the pulse. The variable x establishes the position of the pulse, and the variable t establishes the time. To investigate the properties of solitons, first set the value of constant c: $c = 2$.

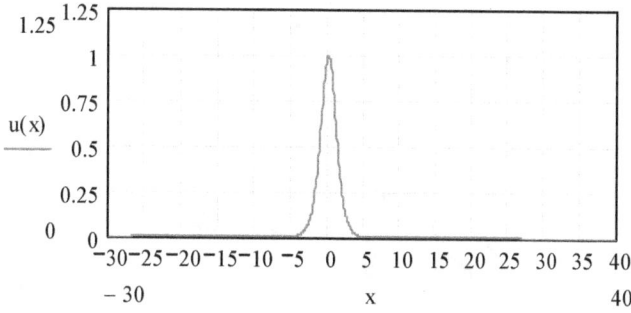

Figure 1.3.15-1 A Soliton Pulse

The soliton pulse moves to the right, while preserving its shape. Since the constant c sets the amplitude and the velocity, the taller the amplitude, the faster the soliton moves.

---

[46] Chuu Lian-Terng and Karen Uhlenbeck, "Geometry of Solitons", Notices of the AMS, January 2000, pp. 17-25.

Note also that since everything is specified in terms of the amplitude, a taller soliton will be skinny whereas a shorter soliton will be fatter. This differs significantly from a monocycle, in that a monocycle's aspect ratio does not change with amplitude, nor does its velocity change as a function of amplitude.

In spite of these differences between solitons and monocycles, UWB Monocycles based on perfect Gaussian pulses exhibit waveshapes similar to solitons. Because of this, solitons and monocycles exhibit many similar properties in terms of their energy-spreading properties, with both spreading energy over ultra wide bandwidths. The biggest differences are that UWB Monocycles are based on truncated Gaussian series, and it is possible to modulate the position of the monocycle pulses to transmit information. The effect on the time waveform of the monocycle from truncation of the ideal Gaussian series terms, as discussed previously, is to add negative undershoot to both the leading and trailing edges of the monocycle pulse. (Solitons can also transmit information, but their inability to support shifting the position as required in PPM limits their usefulness in impulse radio operation.) Still, there is a variation on the regular monocycle that can increase the emission of energy of UWB transmitters at frequencies below the peak of the UWB emissions, where this is desirable, that can be obtained through using soliton concepts.

The Taylor Series expansion for the hyperbolic secant is given by:[47]

$$1 - \frac{x^2}{2} + \frac{5 \cdot x^4}{24} - \frac{61 \cdot x^6}{720} + \ .. + \frac{(-1)^n}{2n!} \cdot \frac{En \cdot x^{2n}}{(2n)!} + \ ..$$

where

$$|x| < \frac{\pi}{2}$$

and En is an Euler Number, i.e., 1, 5, 61, 1385, etc.

Consider two possible truncated hyperbolic secant series approximations for a soliton-variant monocycle:

$$pa2(t) := \left[1 - \frac{\left[\frac{(t - \tau d)}{\tau n}\right]^2}{2} + \frac{5 \cdot \left[\frac{(t - \tau d)}{\tau n}\right]^4}{24} - \frac{61 \cdot \left[\frac{(t - \tau d)}{\tau n}\right]^6}{720} \right] \cdot \exp\left[-2\pi \cdot \left(\frac{t - \tau d}{\tau n}\right)^2 \right]$$

$$pa3(t) := \left[1 - \frac{\left[\frac{(t - \tau d)}{\tau n}\right]^2}{2} \right] \cdot \exp\left[-2 \cdot \pi \cdot \left(\frac{t - \tau d}{\tau n}\right)^2 \right]$$

These differ only slightly from the more traditional UWB monocycle waveform that has been discussed at length already in this paper:

$$p(t) := \left[1 - 4 \cdot \pi \cdot \left(\frac{t - \tau d}{\tau n}\right)^2 \right] \exp\left[-2\pi \cdot \left(\frac{t - \tau d}{\tau n}\right)^2 \right]$$

---

[47] eFunda, Engineering Fundamentals, http://www.efunda.com/math/taylor_series/hyperbolic.cfm, Wolfram Research, retrieved 16 June 2003.

Plotting these three waveforms, and comparing their waveforms, considerable similarity is evident between the two soliton-based approximations, and there is also a strong resemblance from the soliton-based approximations and the ideal Gaussian pulse upon which a UWB monocycle is based:

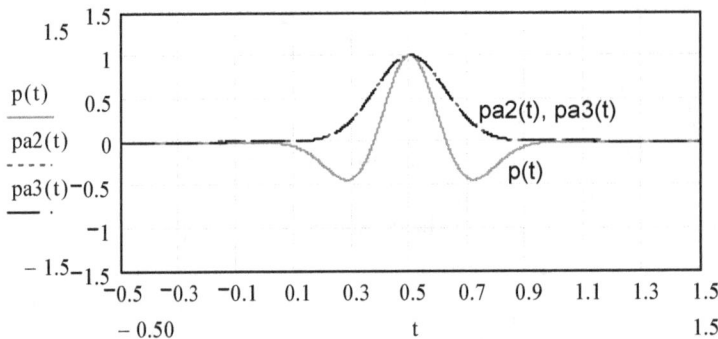

Figure 1.3.15-2 Soliton-based Monocycle Approximations vs. UWB Monocycle

From this plot of the three approximations, it is evident that the soliton-based approximations are much better behaved than the Gaussian pulse approximation in terms of sensitivity to Gibbs-phenomenon caused by series convergence; i.e., the soliton-based approximations retain their fundamental theoretical shapes better with far fewer terms than the Gaussian pulse approximation. Comparing these soliton-variant monocycles with the normal truncated Gaussian monocycle waveform, only the Gaussian monocycle introduces negative undershoot on both the leading and trailing edges. In other words, the penalty is not nearly as great for truncating the approximation series with the soliton-variant monocycle as it is with the conventional Gaussian monocycle.

What is the effect on the power spectrum of using a soliton-variant monocycle? As before, consider a pulse train, using the simplifications identified previously:

$$a := \frac{2 \cdot \pi}{\tau n^2} \qquad T := 3$$

The soliton-variant monocycle function, pa3(t), can be rewritten as svm(t) with these simplifications as:

$$\text{svm}(t) := \left[1 - \frac{a}{4 \cdot \pi} \cdot (t - T)^2 \right] \cdot \exp\left[-a \cdot (t - T)^2 \right]$$

Plotting this function:

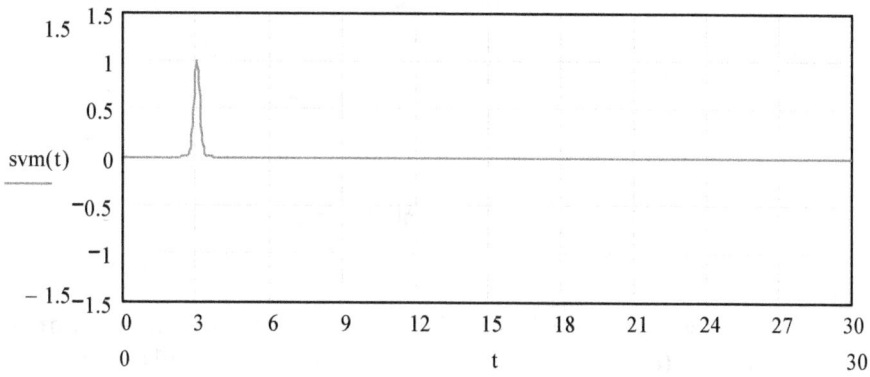

Figure 1.3.15-3 A Single Soliton-variant Monocycle

Note the complete lack of undershoot on this waveform, as compared to a traditional UWB monocycle waveform.

Now, consider a pulse train of 9 of these soliton-variant monocycle pulses, for m = 9:

$$ \text{hsvm}(t) := \sum_{n=1}^{m} \left[\left[1 - \frac{a}{4 \cdot \pi} \cdot (t - n \cdot T)^2 \right] \cdot \exp \left[-a \cdot (t - n \cdot T)^2 \right] \right] $$

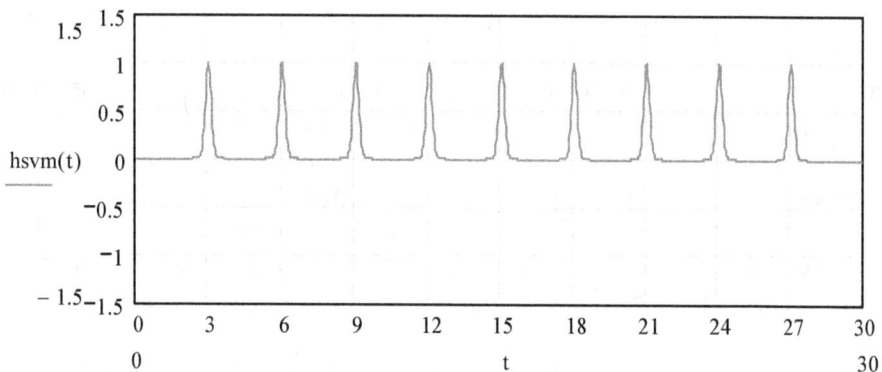

Figure 1.3.15-4 A Soliton-variant Monocycle Pulse Train

As before, this pulse train has no undershoot on the leading and trailing edges of the individual pulses.

The amplitude spectrum is:

$$Fsvm(\omega) := \int_{1.5}^{28.5} hsvm(t) \cdot e^{-2 \cdot \pi \cdot 1i \cdot \omega \cdot t} \, dt$$

and the power spectrum is:

$$Psvm(\omega) := 10 \cdot \log\left(\frac{|Fsvm(\omega) \cdot Fsvm(-1i \cdot \omega)|}{2 \cdot \pi} \right)$$

Plotting the power spectra of the traditional UWB monocycle pulse train, P(ω), against the cosine pulse train, P2(ω), with this soliton-variant monocycle pulse train, Psvm(ω), shows the following:

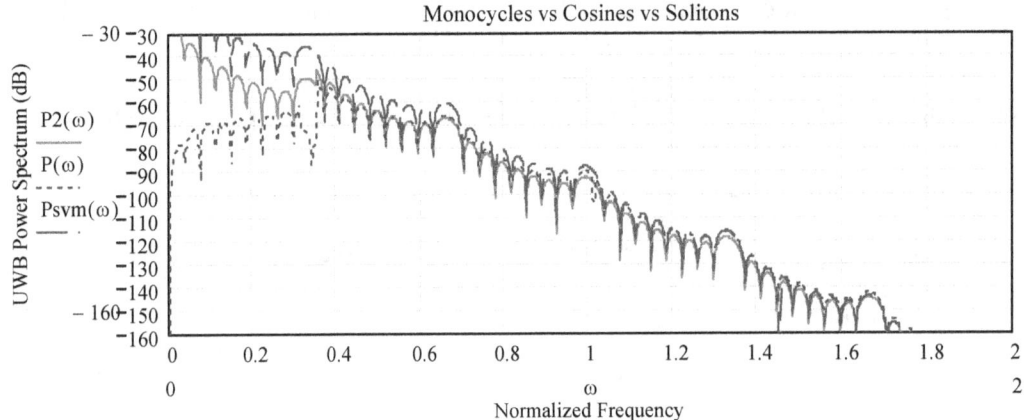

Figure 1.3.15-5 Comparing Power Spectra: SVM vs. Cosine vs. Monocycle

The soliton-variant monocycle power spectrum is hence worse than even the cosine pulse train in not rolling off at lower frequencies below the peak of the power spectrum of the UWB pulse train. The traditional UWB monocycle Gaussian-based pulse train is thus best among the three waveforms for minimizing the effect on narrowband communications below the peak of the output power spectrum. It is therefore the best of the three for reducing interference to GPS receivers operating at frequencies below the FCC Part 15 3.1 GHz to 10.6 GHz UWB band.

On the other hand, for generating test signals for exercising all input frequencies, such as for determining susceptibility of equipment to electronic warfare (EW) jammers, or for even generating EW jamming signals, the soliton-variant monocycle waveform has considerable advantage over the UWB Gaussian-based monocycle as well as over the cosine-based pulse waveforms for covering more frequencies more uniformly with energy in a monotonic power spectrum. Depending on the application, both the traditional UWB Gaussian-based monocycle and the soliton-variant monocycle have clear advantages.

1.3.16   <u>Back to the Future: Damped Sinusoidal Pulses vs. Monocycles</u>

During the earliest days of radio, all transmitters were broadband emitters. Prior to about 1918, the only limit to the bandwidth of a transmitter was often the operating bandwidth of the transmitting antenna (i.e., the aerial) itself.[48] Broadband sparkgap transmitters based on this earliest sparkgap technology ruled the airwaves prior to 1918, and more advanced methods to restrict transmitter emissions (transmissions) to a narrower range of frequencies did not come into widespread use until CW (Continuous Wave) transmitters became dominant around 1920-1921. From about 1918 until 1921, during the last years of "King Spark" as the modulation was often called, limiting the emissions of sparkgap transmitters through using damped sinusoids (also known as resonant sparkgap transmissions) was tried in an attempt to save sparkgap equipment producing companies. These "narrowband" sparkgap transmissions consisted of keyed pulses of decaying sinusoids used in place of keyed pulses of equal amplitude sinusoids. A keyed damped sinusoid consisted of a waveform, kds(t), of the form:[49]

Constants:

$$\omega c := 1.5$$

$$As := 1.2$$

$$\lambda := 0.2$$

$$kds(t) := \left(As \cdot e^{-\lambda \cdot t} \cdot \sin(2 \cdot \pi \cdot \omega c \cdot t) \right)$$

---

[48] This limit was also the only limit for many early, first-generation UWB transmitters. It is only with the introduction of the FCC Part 15 limits of 3.1 GHz to 10.6 GHz as of February 14, 2002 that this practice has been legislated away for UWB transmitters.

[49] The choice of the damped sinusoid constants are based on the values recommended in private notes from Time-Domain Corporation (in their Appendix G attachment supplied to the FCC for obtaining approval of their UWB waveform), which were also recommended by Paul Withington of Time-Domain Corporation to the author.

Keying this waveform, for example, to send an "e" in International Morse Code (a single "dit") at a center frequency of 1.5, the time waveform of the damped sinusoid would be:

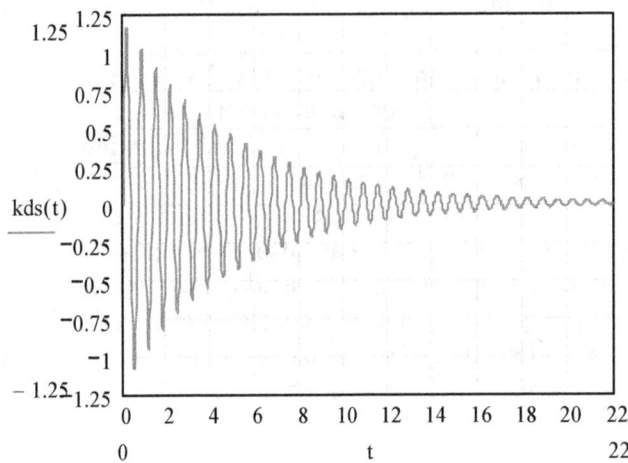

Figure 1.3.16-1 Damped Sinusoid (Class B) Waveform

The rate of decay of this waveform was of considerable interest historically, during the early 1920's, for regulatory reasons. Examining the number of cycles to fall to about 1% of the maximum amplitude:

kds(0.16) = 1.15991

kds(7.5) = 0.26776

$$\frac{kds(21.5)}{kds(0.16)} \cdot 100 = 1.40375$$

Hence, there are approximately 33 oscillations required to fall to about 1% of the maximum amplitude.

During the 1920's, the maximum decrement (decay rate) for resonant spark gap transmitters was set by law not to exceed 0.2. Decrement, δsg, was commonly estimated from simply counting the number of oscillations that occurred for a damped sinusoid to fall to about 1% of the maximum amplitude, and by then using the following equations:[50]

$Noscs := 33$

$$\delta sg := \frac{4.6}{Noscs - 1}$$

$$\delta sg = 0.14375$$

This example, with its estimated decrement factor of 0.14, is less than the 0.20 that was the statutory limit, and so this transmission would have been entirely legal under the early 1920's US radio law.

But, why is this decrement equation what it is, and why does this simple equation work at all? To see this, consider the following:

$$e^{-\lambda \cdot t\_1\%} = 0.01$$

Solving for the t_1% variable:

$$t\_1\% := \frac{4.605170185988091368}{\lambda}$$

The period of the decaying oscillation, of course, is just $2\pi/2\pi\omega c$. (Note: $2\pi/b$ is the period of a sinusoid of form sin (bt).) The approximate number of periods to reach t_1% would be:

$$n\_osc = \frac{t\_1\%}{\frac{1}{\omega c}}$$

---

[50] John H. Morecroft, *Elements of Radio Communication*, John Wiley & Sons, Inc., New York, NY, 1929, pp. 156-162.

First, normalize ωc to unity (convenient enough for examining an unscaled graph). Next, there is a quarter-cycle offset in the damped peaks before reaching the first peak starting at zero time that must come from some of the t_1% elapsed time. To approximately account for this, we just count one oscillation peak less.

$$n\_osc - 1 = t\_1\%$$

$$n\_osc - 1 = \frac{4.605170185988091368}{\lambda}$$

$$\lambda = \frac{4.605170185988091368}{n\_osc - 1.}$$

Now, λ is seen to be the same as the decrement factor, δsg, that was in the quoted equation (taken from the 1929 reference) that was used to estimate the decrement factor by simple counting the number of oscillations required for a resonant spark gap signal to drop to about 1% of its peak amplitude.

From this simplistic derivation, the ancient decrement-estimating equation is seen to give an estimate of the decrement factor while being based on valid mathematical principals. It also appears to have effectively allowed a slight "fudge-factor" in permitting a resonant spark gap transmitter's decrement to exceed the then statutory 0.2 in that it estimates an actual decrement of 0.2 as being only 0.14. Still, approximately 80 years ago when this equation would have been used, it did manage to simplify what would otherwise have been a difficult measurement into an easily-observed oscillograph plot measurement.

What are the spectral emissions for this resonant sparkgap antique radio waveform? The spectrum of the amplitude of one damped sinusoidal pulse can be found as follows.

$$kds(t) := \left(As \cdot e^{-\lambda \cdot t} \cdot \sin(2 \cdot \pi \cdot \omega c \cdot t) \right)$$

$$Fkds(\omega) := \int_0^4 kds(t) \cdot e^{-2 \cdot \pi \cdot 1i \cdot \omega \cdot t} \, dt$$

$$Pkds(\omega) := 10 \cdot \log\left(\frac{|Fkds(\omega) \cdot Fkds(-1i \cdot \omega)|}{2 \cdot \pi} \right)$$

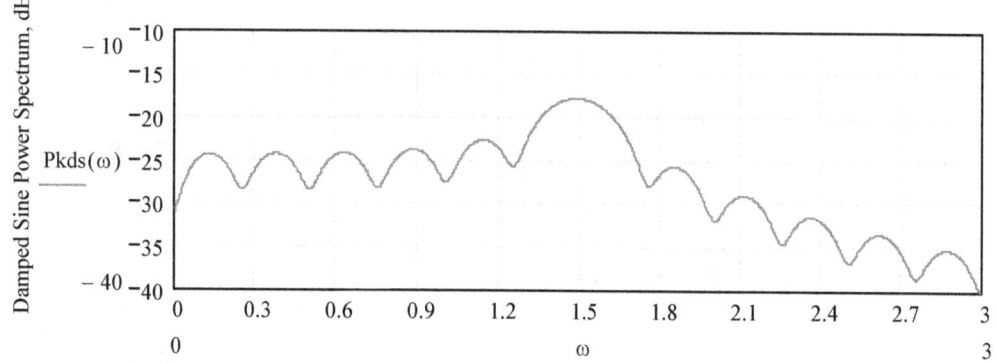

Figure 1.3.16-2 Damped Sinusoid Waveform Power Spectrum

Note the peak at $\omega c = 1.5$.

If preferred, a more classic approach can be applied through using Fourier Transform theory multiplication principals. Consider first the Fourier transform of the one-sided exponential, $e^{\wedge}(-\lambda t)$; its Fourier transform is found as:

$$\left(\int_0^\infty e^{-\lambda \cdot t} \cdot e^{-2 \cdot \pi \cdot i \cdot \omega \cdot t}\, dt \right) \rightarrow \lim_{t \to \infty} \left[\frac{1}{-\lambda - 2 \cdot i \cdot \pi \cdot \omega} \cdot \exp\left[-t \cdot (\lambda + 2 \cdot i \cdot \pi \cdot \omega)\right] + \frac{1}{\lambda + 2 \cdot i \cdot \pi \cdot \omega} \right]$$

The Fourier Transform of a one-sided exponential is hence:

$$F\left(As \cdot e^{-\lambda \cdot t}\right) = \frac{As}{\lambda + 2 \cdot 1i \cdot \pi \cdot \omega}$$

That is:

$$F(\omega) := \frac{As}{\lambda + 2 \cdot 1i \cdot \pi \cdot \omega}$$

The Fourier Shifting Theorem enables finding the Fourier Transform of a product of a function, f(t), which has its own Fourier Transform, F(ω), and a sinusoid as follows:

$$F(f(t) \cdot \sin(2 \cdot \pi \cdot \omega c \cdot t)) = 1i \cdot \frac{(F(\omega + \omega c) - F(\omega - \omega c))}{2}$$

Hence, the Fourier Transform of the damped sinusoid can be found as:

$$F\left(A \cdot e^{-\lambda \cdot t} \cdot \sin(2 \cdot \pi \cdot \omega c \cdot t)\right) = 1i \cdot \frac{(F(\omega + \omega c) - F(\omega - \omega c))}{2}$$

$$1i \cdot \frac{(F(\omega + \omega c) - F(\omega - \omega c))}{2} \rightarrow 1i \cdot \left(\frac{1}{2} \cdot F(\omega + \omega c) - \frac{1}{2} \cdot F(\omega - \omega c) \right)$$

$$\frac{1}{2} \cdot 1i \left(F(\omega + \omega c) - F(\omega - \omega c) \right) \rightarrow \frac{1}{2} \cdot 1i \left[\frac{1.2}{\lambda + 2 \cdot 1i \pi \cdot (\omega + \omega c)} - \frac{1.2}{\lambda + 2 \cdot 1i \pi \cdot (\omega - \omega c)} \right]$$

$$Fds(\omega) := \frac{1}{2} \cdot 1i \left[\frac{1.2}{\lambda + 2 \cdot 1i \cdot \pi \cdot (\omega + \omega c)} - \frac{1.2}{\lambda + 2 \cdot 1i \cdot \pi \cdot (\omega - \omega c)} \right]$$

$$Pds(\omega) := 10 \cdot \log \left(\frac{|Fds(\omega) \cdot Fds(-1i \cdot \omega)|}{2 \cdot \pi} \right.$$

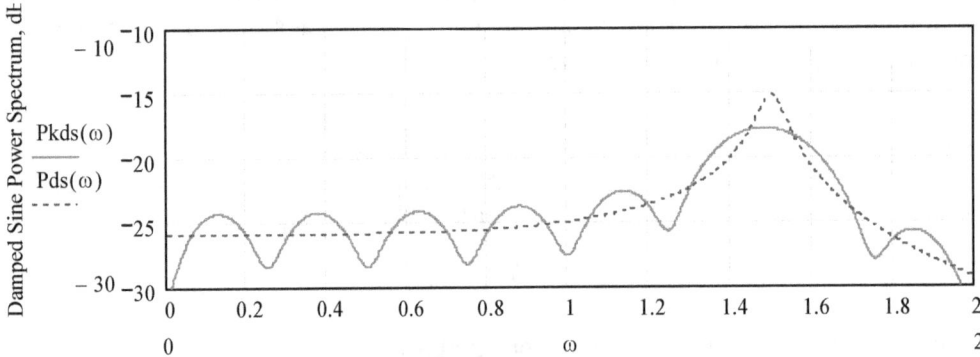

Figure 1.3.16-3 Damped Sinusoid Waveform Power Spectra

For faster decrements, e.g., for λ around 1.1, the power spectra predicted for single damped sinusoidal pulses would become exactly the same using either approach. For slow ringing, 1920's legal, decrements, as shown here, a classic Fourier Transform analysis approach only approximates the actual "bouncing" spectral occupancy shown here.

This slight reduction in spectrum occupancy was a first step in limiting the bandwidth of early spark gap transmitters and was achieved through applying damped sinusoid transmitter technology, also known as resonant sparkgap transmitter technology.

Unfortunately, the reduction in spectral occupancy was only a fraction of what was needed to ease significantly, let alone eliminate, interference among a multitude of resonant spark gap transmitters. (At the time, this was actually billed as a feature; it would be easier for ships-at-sea to attract attention from other ships when sending distress signals, thereby jamming their communications, when using this waveform!)

The ultimate outcome was to categorize this waveform as a Class B Damped Sinusoid, and, by international agreement, to ban forever this waveform from ever being used again. This prohibition still exists, in Regulation 47 CFR, Section 15.5 (d) that states, "Intentional radiators that produce Class B emissions (damped wave) are prohibited."[51]

UWB transmissions are clearly different than damped wave emissions, even though their mathematical waveforms resemble the Class B equation, and this obscure technical section in the Code of Federal Regulations subsequently does not ban modern UWB transmissions.

Now, consider a pulse train of m-pulses of this damped sinusoid, representing a non-PPM pulse train (representing, say, all ONEs or all ZEROs), for m = 9:

$$\text{ptkds}(t) := \sum_{n=1}^{m} \left(\Phi(t - n \cdot T) \cdot \text{kds}(t - n \cdot T) \right)$$

The Heaviside function, $\Phi(t\text{-}nT)$, simply eliminates the problem of the exponential damping 'blowing up' for negative time. Plotting this series of damped sinusoidal pulses:

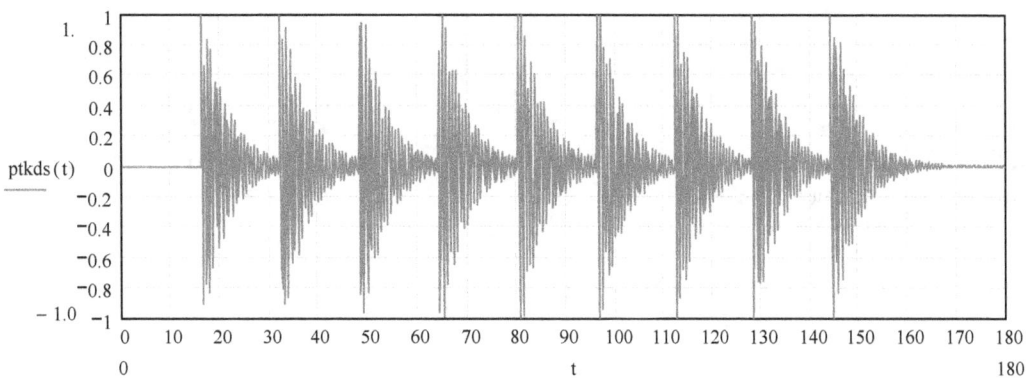

Figure 1.3.16-4 Damped Sinusoid Pulse Train

Unfortunately, the Heaviside Function cannot be integrated across where (t-nT) equals zero. This means that the integration process used previously for finding the power spectra cannot be used for this pulse train. Clearly, though, its spectrum would not be benign to narrowband emissions, nor is it even permitted under current international radio law that bans Class B emissions.

1.3.17   Continuous Wavelet Transforms

The immediate question is why a discussion on Continuous Wavelet Transforms (CWTs) belongs in a discussion that is otherwise focused on UWB monocycles. The reason is that there is a very strong mathematical connection between CWT enabling wavelets and monocycles that has largely gone unnoticed in the literature. Because of this strong

---

[51] Code of Federal Regulations (CFR), 47CFR, Ch. I, (10-1-02 Edition), pp. 678-679, http://www.gpoaccess.gov/cfr/index.html , retrieved 25 June 2003.

relationship, many of the theoretical details derived for Wavelets and CWTs can be applied directly to current UWB research activities. It is therefore advantageous to include an overview of wavelets in a discussion on monocycles.

The continuous wavelet transform (CWT) that has been used extensively in seismic research, focused on detecting signals comprised of transient bursts, is based on the *Mexican Hat Wavelet.*[52] This wavelet is fundamentally nothing more than an alternate form of a UWB monocycle waveform.

The *mother wavelet*, or *analyzing wavelet*, for the Mexican Hat CWT is defined by $\Psi(x)$, where:

$$\Psi(x) := \frac{2 \cdot \pi}{\sqrt{w}} \cdot \left[1 - 2\pi \left(\frac{x}{w} \right)^2 \right] \cdot e^{-\pi \cdot \left(\frac{x}{w} \right)^2}$$

and where the wavelet width-parameter, *w*, analogous to the normalizing value, τn, for a monocycle, is, for the following example, set to:

$$w := \frac{1}{8}$$

Define next the maximum wavelet scale parameter, *k*. This value is also called the number of *voices* per octave.

$$k := 2$$

For

$$s0 := 2^{\frac{-0}{k}}$$

it follows:

$$\Psi s0(x) := \frac{1}{\sqrt{s0}} \cdot \Psi \left(\frac{x}{s0} \right)$$

This is really just the *mother wavelet*, $\Psi(x)$.

---

[52] James S. Walker, *A Primer on WAVELETS and their Scientific Applications*, Chapman & Hall/CRC, 1999, p. 130.

Similarly, it follows that:

$$s1 := 2^{\frac{-1}{k}} \qquad \Psi s1(x) := \frac{1}{\sqrt{s1}} \cdot \Psi\left(\frac{x}{s1}\right)$$

$$s2 := 2^{\frac{-2}{k}} \qquad \Psi s2(x) := \frac{1}{\sqrt{s2}} \cdot \Psi\left(\frac{x}{s2}\right)$$

$$s3 := 2^{\frac{-3}{k}} \qquad \Psi s3(x) := \frac{1}{\sqrt{s3}} \cdot \Psi\left(\frac{x}{s3}\right)$$

$$s4 := 2^{\frac{-4}{k}} \qquad \Psi s4(x) := \frac{1}{\sqrt{s4}} \cdot \Psi\left(\frac{x}{s4}\right)$$

$$s5 := 2^{\frac{-5}{k}} \qquad \Psi s5(x) := \frac{1}{\sqrt{s5}} \cdot \Psi\left(\frac{x}{s5}\right)$$

$$s6 := 2^{\frac{-6}{k}} \qquad \Psi s6(x) := \frac{1}{\sqrt{s6}} \cdot \Psi\left(\frac{x}{s6}\right)$$

Since there are $I * k = 6$ scaled wavelets, and $I = 3$ for $k = 2$, we say that there are 3 octaves in this finite collection of scale parameters. "I" is simply the number of octaves.

Plotting this set of scaled wavelets:

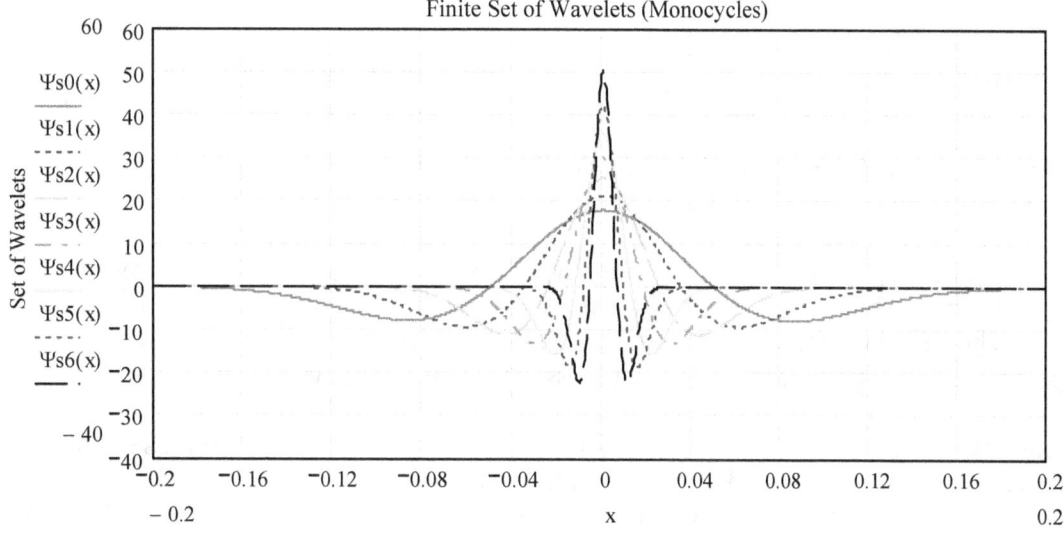

Figure 1.3.17-1 Scaled Wavelets are scaled from the Mother Wavelet

The value of a CWT is that it enables a finely detailed frequency analysis of a signal waveform in both time *and* frequency, unlike a Fourier Transform that enables a somewhat haphazardly performed frequency analysis, only. [Note: For the case of UWB monocycles, time and frequency are exactly the parameters that matter most for determining modulation information (i.e., the data content).] A CWT of a function consists of a finite collection of correlations done with a finite collection of scaled wavelets (i.e., scaled monocycles.)

Wavelets used for detection purposes, such as in seismic research, exhibit *M* vanishing moments. That is to say, the mathematical property for a *mother wavelet* exhibits the following property when used to detect signals that contain transient bursts:

$$\int_{-\infty}^{\infty} \Psi(x) \cdot x^n \, dx = 0$$

for $\quad n = 0..(M-1)$

Does the Mexican Hat Wavelet exhibit this mathematical property?

$$\int_{-\infty}^{\infty} \Psi(x) \cdot x^0 \, dx = 0.00000 \qquad \int_{-\infty}^{\infty} \Psi(x) \cdot x^1 \, dx = 0.00000 \qquad \int_{-\infty}^{\infty} \Psi(x) \cdot x^2 \, dx = -0.01105$$

$$\int_{-\infty}^{\infty} \Psi(x) \cdot x^3 \, dx = 0.00000 \qquad \int_{-\infty}^{\infty} \Psi(x) \cdot x^4 \, dx = -0.00016 \qquad \int_{-\infty}^{\infty} \Psi(x) \cdot x^5 \, dx = 0.00000$$

$$\int_{-\infty}^{\infty} \Psi(x) \cdot x^6 \, dx = -0.00000$$

$+$

Neglecting roundoff error, the Mexican Hat Wavelet therefore does exhibit vanishing moments.

Why is this property useful? It has importance because it enables ignoring lower-order signal bursts, while instead focusing on detecting just higher-order signal bursts. This works because integrating a Mexican Hat Wavelet of degree M with a mathematical polynomial (i.e., a signal) of a degree of less than, or equal to, M - 1 results in a zero.

However, around the specific times that a polynomial (i.e., the signal) exhibits higher-order variations, namely of degree M or larger, the result is not zero. Because of this, a Mexican Hat Wavelet of an appropriate order is well suited to detecting and successfully flagging the occurrence of transient signal bursts containing higher-order degrees of variations. Unlike a Fourier Transform, which is often unable to detect a transient burst, a CWT such as a Mexican Hat Wavelet can instead successfully detect such transitory events.

In a traditional Fourier analysis of a function, performed with a Fourier Transform, only frequency data is considered. For the case of time-domain data containing outliers, or other unusual anomalies, the classic difficulty is always determining just what time scale should be used for performing the Fourier transform integration over the function (or data). With a CWT, the time span selection process is greatly simplified, since the transform handles both the frequency and time scaling difficulties simultaneously. The end result is that a CWT is well suited for analyzing signals containing transient bursts. A CWT is also well suited for analyzing UWB radio communications, and for detecting clandestine or covert UWB transmissions utilizing monocycles having unknown parameters.

To show that detection of unauthorized, clandestine, covert UWB transmissions can be accomplished in a detection process by using a Continuous Wavelet Transform, consider a covert UWB transmission monocycle pulse having unknown, modified parameters of the form:

$$ p\_unk(t) := \left[1 - 4 \cdot \pi \cdot \left(\frac{t - \tau d}{\tau n \cdot 0.5} \right)^2 \right] \exp\left[-2\pi \cdot \left(\frac{t - \tau d}{\tau n \cdot 0.5} \right)^2 \right] $$

Now, let us apply a series of scaled wavelets in a CWT and see if this covert UWB signal can be detected. Starting with the shortest (largest scale factored) wavelet, scaled, of course, from the *mother wavelet*:

$$ \Omega s6(\tau) := \int_{-\infty}^{\infty} p\_unk(t) \cdot \Psi s6(t - \tau) \, dt $$

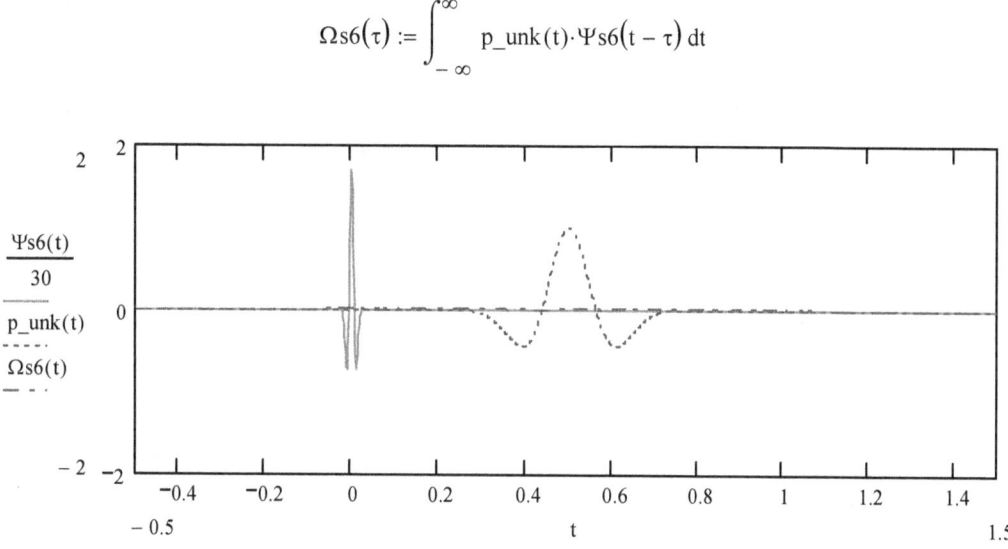

Figure 1.3.17-2 First Scaled Wavelet Detects Nothing

Clearly, nothing was detected.

Trying again, with the next scaled wavelet:

$$\Omega s5(\tau) := \int_{-\infty}^{\infty} p\_unk(t) \cdot \Psi s5(t - \tau)\, dt$$

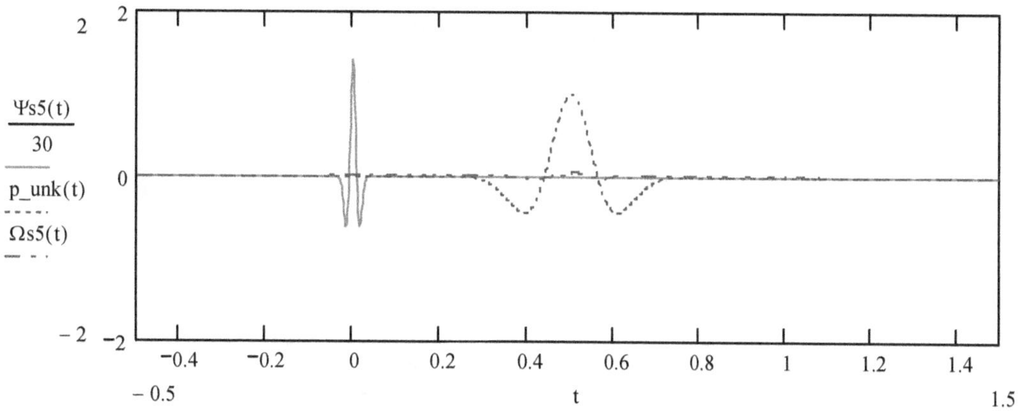

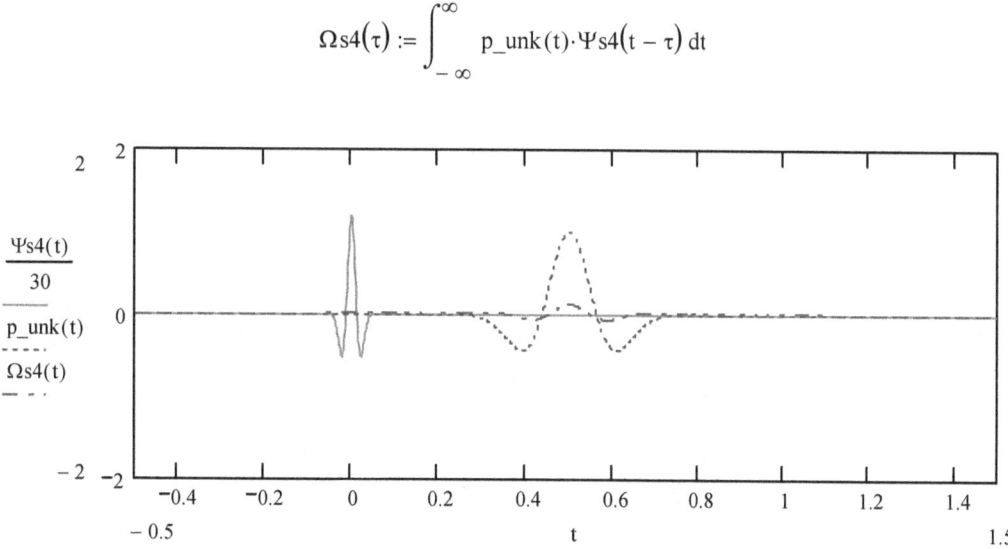

Figure 1.3.17-3 Second Scaled Wavelet Still Detects Nothing

Likewise, there is no clear indication of the presence of a monocycle.

Continuing to the next scaled wavelet:

$$\Omega s4(\tau) := \int_{-\infty}^{\infty} p\_unk(t) \cdot \Psi s4(t - \tau)\, dt$$

Figure 1.3.17-4 Third Scaled Wavelet Hints of Detection

There is a hint of a monocycle starting to be detected.

Continuing to the next scaled wavelet:

$$\Omega s3(\tau) := \int_{-\infty}^{\infty} p\_unk(t) \cdot \Psi s3(t - \tau)\, dt$$

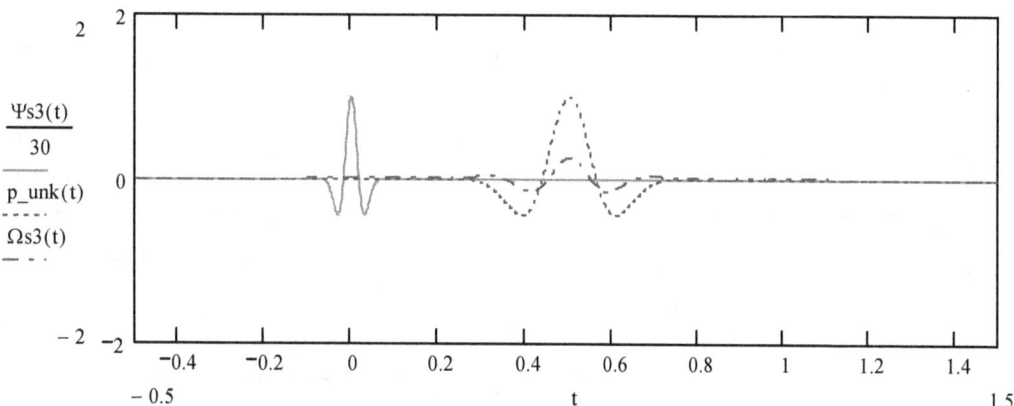

Figure 1.3.17-5 Fourth Scaled Wavelet Shows Evidence of Detection

There is clearly now evidence that a monocycle is present.

Continuing to the next scaled wavelet:

$$\Omega s2(\tau) := \int_{-\infty}^{\infty} p\_unk(t) \cdot \Psi s2(t - \tau)\, dt$$

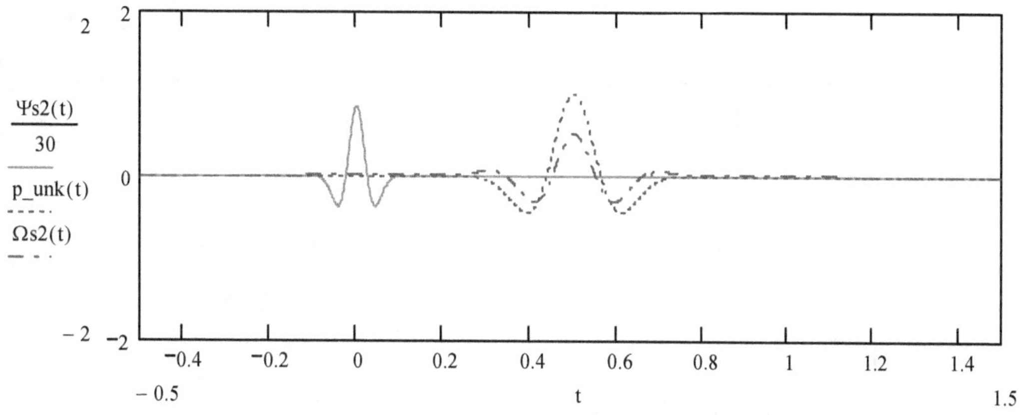

Figure 1.3.17-6 Fifth Scaled Wavelet Shows Clear Evidence of Detection

The monocycle is clearly showing its presence now, and is growing.

Continuing to the next scaled wavelet:

$$\Omega s1(\tau) := \int_{-\infty}^{\infty} p\_unk(t) \cdot \Psi s1(t - \tau)\, dt$$

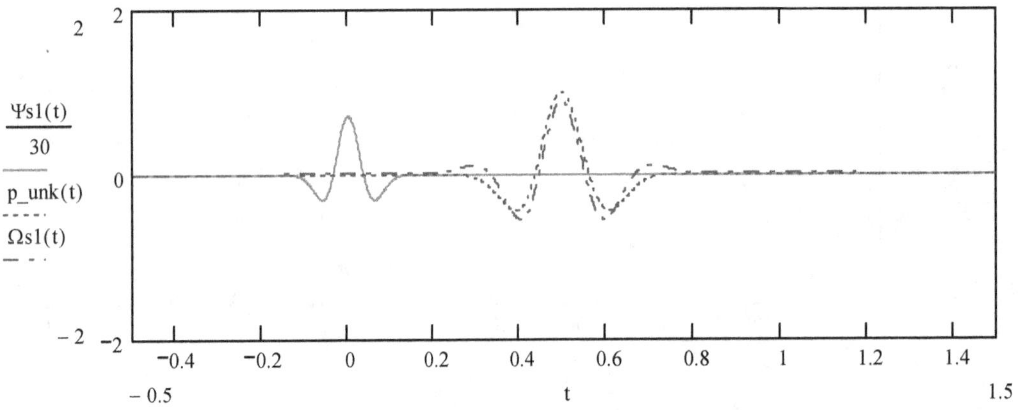

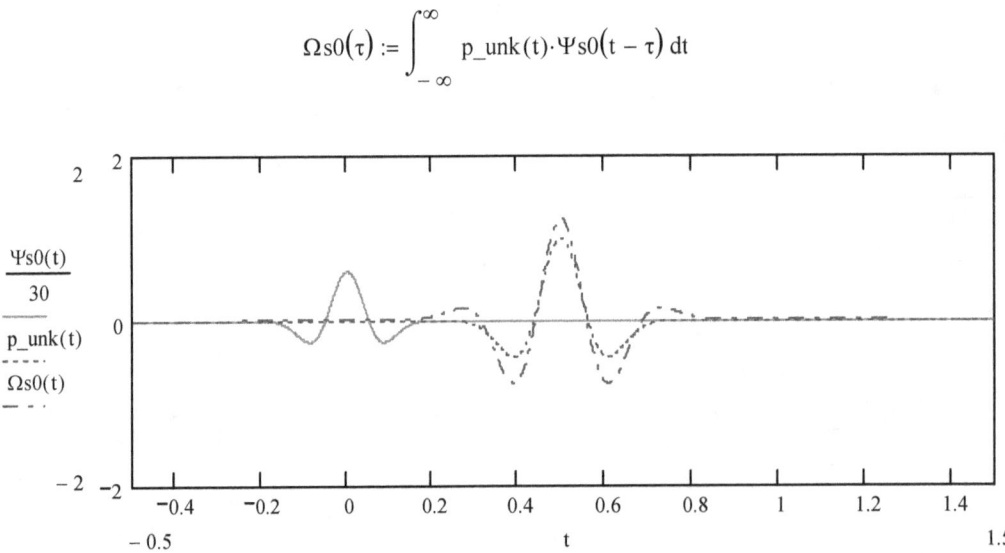

Figure 1.3.17-7 Sixth Scaled Wavelet Clearly Detects Monocycle

The detected monocycle, detected without having to know the details of the monocycle, is approaching the magnitude of the received monocycle.

Continuing to the next scaled wavelet:

$$\Omega s0(\tau) := \int_{-\infty}^{\infty} p\_unk(t) \cdot \Psi s0(t - \tau)\, dt$$

Figure 1.3.17-8 Seventh Scaled Wavelet Detects Similar to a Matched Receiver

A clandestine UWB monocycle is definitely present, and has been detected equally well as if by a UWB receiver having had the good fortune of knowing in advance all of the parameters of the transmitted clandestine UWB monocycle.

Without assuming any particular parametric values for the clandestine UWB monocycle, it is therefore possible to detect the presence of a monocycle through using CWT techniques, by transforming a sampled signal with a set of scaled wavelets spanning multiple octaves. This is important, for with conventional Fourier analysis spectrum analyzer techniques, it is difficult to detect the presence of a UWB monocycle even when knowing that it is present. [Based on measurements in the lab, a maximum detection distance of only a few inches was the limit at which a spectrum analyzer with an external Low Noise Amplifier and antenna could detect a UWB transmission from a UWB transmitter antenna.] With a Continuous Wavelet Transform technique, it is thus possible to increase the detection distance to a value approaching the maximum communication range of the UWB link designed *a priori* with complete knowledge of the monocycle waveform chosen.

A standard figure of merit for clandestine transmitters can be defined in terms of the ratio of maximum communication range to the maximum detection range, assuming no details are known about the signal. For example, if a specific UWB transmitter has a range of 150 feet when communicating with a receiver designed based on known UWB monocycle parameters, and the maximum detection distance is only 1/4 foot for the uncooperative case, then the Figure of Merit for Covertness would be 150/(0.25), or 600. With a CWT technique, it would be possible to reduce this Figure of Merit of Covertness to perhaps 1.5, for an improvement of over 52 dB versus a prior-art, non-CWT technique.

With this much improvement in detection, a CWT-detection technique has considerable value for ferreting out clandestine UWB transmitters. In all likelihood, the detection bubble would likely be 2/3 or more of the communication range bubble, greatly reducing the volume (and area) over which a security sweep would need to be conducted to guarantee detection of a covert UWB transmitter with a high probability of intercept (POI). The only significant impediment with this technique is that the CWT hardware would need to employ a wide-range of wavelet widths, and the maximum sample rate would need to be at least twice the maximum frequency occupied by the UWB signal to guarantee finding the unknown UWB monocycle transmission. The detection equipment would also be slightly more complex than the typical UWB receiver designed with *a priori* knowledge of the monocycle particulars. Still, this would not be unreasonable, since it is likely that only a small number of such detection units would be required for sweeping areas securely, while guaranteeing that covert UWB transmitters were not in operation.

1.3.18   Third-Order Introduction to Monocycles

Classic electromagnetics theory historically has always been applied to designing antennas while tacitly assuming steady-state responses. The reason for making this simplifying assumption is that it greatly simplifies the application of Maxwell's Equations for designing antennas and simultaneously permits using simplified Electromagnetics Theory, thereby avoiding the discontinuous anomalies that exist during the startup of the waveform (i.e., antenna capacitance charging effects.)

For UWB antennas, there is no shortcut that can be used to avoid the discontinuous events at the start of the pulse waveform, for this is all there is. Steady-state electrical conditions are never reached in UWB antennas. (Although, arguably, steady-state heating effects for high-power transmissions are reached, on an average basis.) The need to include complete, un-simplified Maxwell Equations in designs is further exacerbated by the need for designing ultra wide bandwidth antennas to be compatible with transmitter output amplifiers, to guarantee unconditional stability in power amplifier output circuits. (Fortunately, there are analysis suites of modeling tools available that can do the time-domain analysis with Maxwell's Equations without assuming any steady-state approximations, e.g., commercial 3D Electromagnetic Simulation products such as IE3D, and, to a lesser extent, HFSS.)[53,54]

The main issue, of course, is that, during the initial rush of UWB monocycle current into an antenna, the antenna acts as an open-circuit and must be charged. The effect, however, is that the current in the antenna structure is phase-shifted by 90 degrees, which means that the UWB monocycle current pulse input into the structure has its derivative taken. Then, upon the Electromagnetic Wave impinging on a receiving antenna, the same derivative operation occurs again. The received signal is therefore the scaled first derivative of the Electric Field resulting from the original UWB current pulse, and is the scaled second derivative of the original UWB current pulse. Although this summarizes what happens, a more concise mathematical explanation is in order. From basic antenna theory, the electric field radiated from the antenna (i.e., the E-field, E(t)) is proportional to the derivative of the magnetic potential, A(t). That is:[55]

$$E(t) = k1 \cdot \left(\frac{d}{dt} A(t) \right)$$

However, the magnetic potential is proportional to the current flow in the antenna structure:

$$A(t) = k2 \cdot i(t)$$

The E-field radiated from an antenna is therefore proportional to the first derivative of the current flow into the antenna. For sinusoidal currents, the taking of the derivative of the

---

[53]IE3D is available from Zeeland Software, Inc., http://www.zeland.com/

[54] HFSS is available from Ansoft Corporation, http://www.ansoft.com/products/hf/hfss/index.cfm

[55] Michael Chia, *"UWB Radio for wireless communications - I2R's perspectives,"* Ultra Wideband (UWB) Programme, Singapore Suntec Convention Center, Singapore, 25 February 2003 (an IDA UWB Seminar).

sinusoidal current becomes a co-sinusoidal current, or, equivalently, a phase-shifted sinusoidal current. The derivative process is therefore generally ignored for continuous wave signals, being equivalent to simply a shift in the apparent position of the original transmitter antenna position.

For non-sinusoidal pulses, however, such as are used for UWB transmissions, the derivative taking process becomes entirely different, causing the waveform to change its shape fundamentally. A transmitted monocycle hence will appear different, depending on where the monocycle is observed, and how it is observed, unlike a sinusoidal waveform.

Gaussian current pulses and signal pulses partially preserve their shapes when their derivatives are taken, at least for perfectly shaped Gaussian pulses. For truncated Gaussian approximations, however, such as occur in actual UWB radio implementations, the perfect shapes are not completely preserved when their derivatives are taken. Likewise, the computed power spectrums are not the same for the different derivatives, either, further exacerbating issues such as meeting FCC spectral masks imposed on UWB transmissions. Furthermore, true Gaussian pulses technically exhibit an infinite pulse-width. The trick that is most commonly used to overcome the infinite pulsewidth issue is to define UWB monocycles and Gaussian pulses as having a defined, although finite, pulsewidth that contains, say, 99.99% of the energy of the theoretical Gaussian pulse.

Unlike the series approximation to a Gaussian pulse used previously, a non-series, closed-form approximation to a Gaussian pulse can be written more compactly as follows:[56]

$$p\_gp1(t) := \frac{A}{\sqrt{2\cdot\pi\cdot\sigma}}\cdot\exp\left(\frac{-t^2}{2\cdot\sigma^2}\right)$$

This Gaussian pulse will be considered to be the UWB monocycle current pulse into the transmit antenna. The electromagnetic field from the transmitter antenna is then related to the scaled first derivative of this current pulse, that is, to:

$$\frac{d}{dt}\left(\frac{A}{\sqrt{2\cdot\pi\cdot\sigma}}\cdot\exp\left(\frac{-t^2}{2\cdot\sigma^2}\right)\right)$$

---

[56] Hongsan Sheng, Philip Orlik, Alexander M. Haimovich, Leonard J. Cimini Jr., Jinyun Zhang, *"On the Spectral and Power Requirements for Ultra-Wideband Transmission,"* IEEE International Conference on Communications, Anchorage, AK, May 2003.

Evaluating this:

$$\frac{d}{dt}\left(\frac{A}{\sqrt{2\cdot\pi}\cdot\sigma}\cdot\exp\left(\frac{-t^2}{2\cdot\sigma^2}\right)\right) \rightarrow \frac{-1}{2}\cdot A\cdot\frac{2^{\frac{1}{2}}}{\pi^{\frac{1}{2}}\cdot\sigma^3}\cdot t\cdot\exp\left(\frac{-1}{2}\cdot\frac{t^2}{\sigma^2}\right)$$

Hence,

$$p\_gp2(t) := \frac{-1}{2}\cdot A\cdot\frac{2^{\frac{1}{2}}}{\pi^{\frac{1}{2}}\cdot\sigma^3}\cdot t\cdot\exp\left(\frac{-1}{2}\cdot\frac{t^2}{\sigma^2}\right)$$

This Gaussian pulse, p_gp2(t) is what arrives at the receiver antenna as the E-field.

However, the Gaussian current pulse at the output of the receiver antenna is a scaled version of the *derivative of this pulse*; that is:

$$\frac{d}{dt}\left(\frac{-1}{2}\cdot A\cdot\frac{2^{\frac{1}{2}}}{\pi^{\frac{1}{2}}\cdot\sigma^3}\cdot t\cdot\exp\left(\frac{-1}{2}\cdot\frac{t^2}{\sigma^2}\right)\right) \rightarrow \frac{-1}{2}\cdot A\cdot\frac{2^{\frac{1}{2}}}{\pi^{\frac{1}{2}}\cdot\sigma^3}\cdot\exp\left(\frac{-1}{2}\cdot\frac{t^2}{\sigma^2}\right) + \frac{1}{2}\cdot A\cdot\frac{2^{\frac{1}{2}}}{\pi^{\frac{1}{2}}\cdot\sigma^5}\cdot t^2\cdot\exp\left(\frac{-1}{2}\cdot\frac{t^2}{\sigma^2}\right)$$

Hence,

$$p\_gp3(t) := \frac{-1}{2}\cdot A\cdot\frac{2^{\frac{1}{2}}}{\pi^{\frac{1}{2}}\cdot\sigma^3}\cdot\exp\left(\frac{-1}{2}\cdot\frac{t^2}{\sigma^2}\right) + \frac{1}{2}\cdot A\cdot\frac{2^{\frac{1}{2}}}{\pi^{\frac{1}{2}}\cdot\sigma^5}\cdot t^2\cdot\exp\left(\frac{-1}{2}\cdot\frac{t^2}{\sigma^2}\right)$$

To clarify this further, depending on where one observes the UWB monocycle, one may see any of the following waveform shapes, scaled, of course, depending on actual circuit gains, path losses/distances, and actual signal levels:

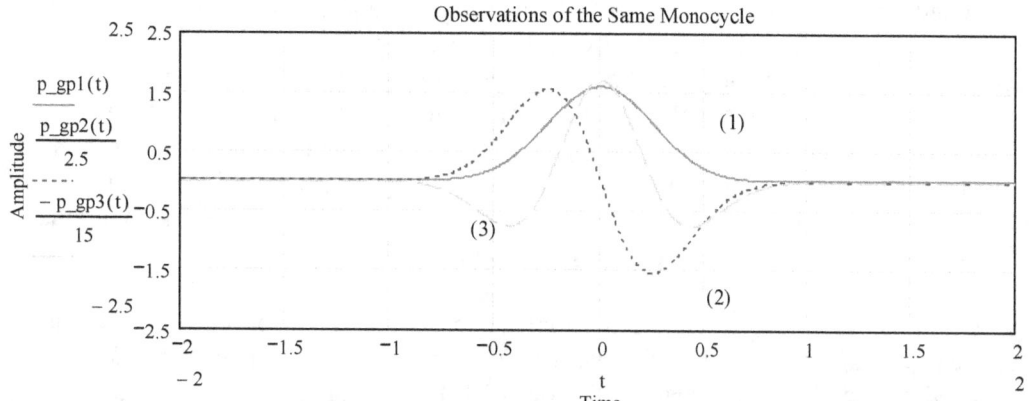

Figure 1.3.18 Observing the same monocycle at different locations

All of these waveforms are observations of the same monocycle, observed at different locations within the UWB communication system. It is important to realize that all of these scaled waveforms are the same UWB monocycle observed as either (1) a current pulse into the transmitter antenna, (2) as an electromagnetic field traveling from the transmitter antenna to the receiver antenna, and (3) as a current pulse out of the receiver antenna. Depending on one's observation point, all of these waveforms are simply a different scaled representation of the *same* UWB monocycle. This can be thought of analogously, perhaps, as a *Theory of UWB Monocycle Relativity*.

An alternative treatment is simply to call these waveforms by different names, i.e., the Gaussian Pulse, the Pulse Doublet (for the first derivative), the Pulse Triplet (for the received current pulse out of the receive antenna, and, if another derivative is taken (the 3rd derivative) as the Pulse Quadlet.[57] (As a mnemonic to remembering these terms, just count the "bumps" on the signal waveform to determine whether to call the monocycle representation a pulse, a doublet, a triplet, or a quadlet.)

All in all, this is rather confusing for many traditional radio designers and antenna designers for, in the steady state, there *is* a derivative being taken, but the waveforms remain invariant, and are only shifted in time. Still, once it is understood that derivatives are taken of UWB Monocycles when passing through antennas, and their time-domain waveform shapes change because of this, the world of UWB Monocycles instantly becomes much more clear.

Unfortunately, however, the power spectra are not the same for all these different observations taken of the same pulse waveform at different points in a UWB

[57] Mark A. Barnes, Soumya K. Nag, Herbert U. Fluher, "*Method of Envelope Detection and Image Generation*," United States Patent US 6,552,677, dated April 22, 2003.

71

communication system. The information content remains the same, of course, but the shape of the pulse changes and the bandwidth occupied by the pulse also changes depending on where one observes the UWB monocycle. From a purely physical point of view, this is difficult to understand. However, if one accepts that there can be localized cancellation of amplitudes in three-dimensional space, much as is seen along a mismatched transmission line where a Voltage Standing Wave Ratio exists where there are nulls in the amplitude at specific physical points along a transmission line, this concept becomes easier to accept. It complicates the issue of just what the power spectrum is of a UWB monocycle-based communication system, however. There may be more than just one power spectrum for a given monocycle that must be considered in terms of interference potential. The factor that ultimately must determine which power spectrum to consider is the point of susceptibility in the susceptible wireless apparatus that may experience interference from the UWB power spectrum.

In general, it is possible to use UWB monocycles based on higher-order derivatives to tailor the power spectral density of transmissions to meet arbitrary spectral masks, such as imposed by the FCC in February 2003 for UWB transmissions.[58] The verification of whether a particular spectral mask is met, however, is an issue still subject to much debate, because different observers of a UWB waveform will see different power spectrums DEPENDING ON WHERE THEIR OBSERVATION POINT IS. This 'Relativity' is a first for radio systems spectrum management.

Much remains before all UWB open issues can be resolved. Ultimately, however, the need to reuse frequency spectrum, and to use frequency spectrum more efficiently, will force the resolution of these details, for the idea of spectral reuse inherent with UWB communications systems holds too great a promise to ignore resolving these details.

1.3.19   Open Theoretical Issues with UWB Communication Systems

Despite the time-synchronization "hand waving" assumed thus far to analyze fundamental aspects of UWB radios, the myriad difficulties surrounding time-synchronization should not be underestimated. The fundamental problems in UWB radio design today can largely be grouped into just one area involving time-synchronization problems, and the associated multiple access issues. The two issues are closely related, as it can be extremely difficult to differentiate between individual monocycles from an assortment of UWB transmitters unless receiver complexity is greatly increased over what is required for implementing a single UWB link.

A common technique used in spread spectrum link designs is a fixed preamble consisting of a training sequence to enable quick recognition of a particular signal, thereby speeding acquisition. With a TH-PPM signal, this could be implemented with a known relative timing Time-Hop sequence of monocycles prepended onto the start of each major

---

[58] Hongsan Sheng, Philip Orlik, Alexander M. Haimovich, Leonard J. Cimini Jr., Jinyun Zhang, *"On the Spectral and Power Requirements for Ultra-Wideband Transmission,"* IEEE International Conference on Communications, Anchorage, AK, May 2003.

transmission (say a packet) that were transmitted using UWB modulation techniques. It would be somewhat the equivalent of a time-domain preamble in place of a frequency domain preamble.

Using a preamble simplifies the correct recognition of a particular signal once monocycle detection occurs, but with low-power UWB monocycle transmissions, monocycle detection often occurs only when timing is acquired. Use of a preamble does not mitigate the basic timing uncertainty problem inherent in UWB receiver designs necessary to enable detection in the first place. It only simplifies recognition of signals from a particular transmitter. As noted by Scholtz, et al, [59] "a one nanosecond time-resolution used in a system with an initial timing uncertainty equivalent to a spreading code period of one millisecond means that the receiver must compute 10^6 correlations. This acquisition problem is easily a few orders of magnitude more difficult than exists for narrowband systems with the same initial uncertainty..."

For expediency in acquiring UWB signals, therefore, the key to success is achieving an efficient parallel correlator architecture to search all the correlator bins quickly and accurately, rather than in a serial fashion, to acquire timing fast. The acquisition problem for UWB transmissions likely poses the largest difficulty in terms of its impact on receiver complexity, power consumption, and physical size of receiver hardware relative to that required with traditional narrowband communication systems. Once timing is acquired, a monocycle pulse train stream can be detected and processed with a complexity and power consumption much less than the hardware required with a typical narrowband system. The fundamental problem is just to obtain proper timing in the first place.

For communication links involving high vehicle velocities with corresponding high rates of distance separation or distance closing, in which Doppler effects becomes prevalent, the timing problem only becomes worse. Spectral frequencies are shifted, pulse phases can change, and timing uncertainties only increase. Doppler effects simply increase the required receiver complexities to even higher levels.

For environments in which multiple UWB signals coexist, the difficulty in distinguishing specific monocycles as to their origin appears intractable without first incorporating higher system level concepts, such as those used in Galois Field computations for determining codeword orthogonality and Hamming distances in Error Correction Coding and spread spectrum spectral orthogonality areas. Of course, for UWB signals, the concepts must be extended to the time-domain, instead of to the codeword polynomial and Walsh function domains, respectively, for the Error Correction Coding and spread spectrum problems.

---

[59] Robert A. Scholtz, P. Vijay Kumar, and Carlos J. Corrada-Bravo, "Signal Design for Ultra-wideband Radio", Sequences and Their Applications (SETA '01), Bergen, Norway, May 13-17, 2001. (Work sponsored by Office of Naval Research under grant N00014-96-1-1192 (subcontract of the Univ. of Puerto Rico), and by the National Science Foundation under grant ANI-9730556.)

Much work remains to make UWB communication techniques practical in specialized environments, especially where multiple UWB transmitters exist and where either receivers or transmitters are moving at high rates of speed relative to each other.

1.4 TESTING DESCRIPTION

Volume II of this Final Report previously described the detailed testing description for the tests conducted on the UWB hardware investigated on this project. For expediency, these descriptions are not repeated here.

1.5 TEST OBJECTIVES

The fundamental objectives for the UWB hardware testing conducted on this project were the interference profiles both from and to the tested UWB hardware relative to conventional wireless (radio) equipment. In order to understand these results, the theoretical results documented earlier in this, Volume III of this Final Report, were derived to enable establishing a firm theoretical understanding of the limits of UWB technology.

1.6 TEST SETUP

Volume II of this Final Report previously described the detailed testing setups used for the tests conducted on the UWB hardware investigated on this project. For expediency, these descriptions are not repeated here.

1.7 TEST EQUIPMENT AND EVALUATION KIT

1.7.1 Test Equipment

Volume II of this Final Report previously described the test equipment used during testing UWB hardware on this project. For expediency, these descriptions are not repeated here.

1.7.2 Evaluation Kit (EVK)

Taking advantage of one of the first commercial products available, an early pair of Evaluation Kit (EVK) TM-UWB radios was purchased and received early in January 2003 from Time-Domain Corporation, of Huntsville, AL. This EVK, consisting of a pair of UWB transmitter/receiver radios with Ethernet link interfaces, along with controlling software for use on a laptop, was the UWB exemplar tested on this project. Working with Time-Domain Corporation, two software firmware upgrades were received, resolving the shortcomings discovered during testing. The testing results of this EVK provide the bulk of the content of the UWB laboratory-testing results, discussed at length in Volume II of this report.

XtremeSpectrum Incorporated (XSI) of Vienna, VA, was also approached regarding the availability of their Bi-phase Pulse Modulation evaluation kit, consisting of a four-chip (now three-chip, e.g., *Trinity^fm*) chipset providing 100 Mb/s data rates and consuming less than 200 mW that was to be priced at only $19.95 in quantities of 100,000.[60] Unfortunately, XSI's Evaluation Kit, originally scheduled for availability by July 2003, has slipped its availability date, and was not available in time for testing on this project.

1.8   HISTORICAL PARALLELS OF ULTRA-WIDEBAND (UWB)

The last disruptive technology shift of a similar magnitude in wireless occurred during World War I. Early wireless signals (i.e., radio signals) were mostly transmitted from 1896 until 1919 with broadband spark-gap transmitters. Capitalizing on this spark-gap transmitter technology, and on his own successful 1896-1898 wireless experiments, Guglielmo Marconi obtained funding in 1899 to found the British Marconi Company. In 1901, he further expanded his company through opening an American subsidiary. With a successful demonstration of communication over the Atlantic Ocean from England to Newfoundland on December 12, 1901, the British Marconi Company became the dominant wireless company in the world. It remained dominant until 1919, when spark-gap radio was replaced with more modern, disruptive technology.

Not all was rosy, however, even during the early years of the British Marconi Company's dominance. The existing trans-Atlantic cable companies and the telephone companies on both sides of the Atlantic were firmly entrenched, and applied considerable business pressure to counter the British Marconi Company's upstart technology. No upstart 'wireless' company would be allowed to threaten the dominant telecom businesses of the day. Because of political pressure from the existing Anglo-American Telephone Company, Marconi left Newfoundland and was forced to re-locate to Nova Scotia. Likewise, the international competition against more robust and established trans-Atlantic cable companies further squeezed Marconi's company through eliminating international trans-Atlantic communication business. The only remaining profitable niche was purely marine business – i.e., the ship-to-ship and the ship-to-shore communication businesses, where, for purely technical reasons, neither of the existing companies could compete; and it was in providing these services where the British Marconi Company found its home.

Spark-gap transmitter signals occupied broad bandwidths for that time, using multiple MHz of bandwidth for sending information theoretically requiring only 100 Hz or less bandwidth for transmission (for an assumed 25 WPM transmission rate). Although wasteful of the RF spectrum, as long as spectral occupancy remained light, this approach

---

[60] Yoshida, Junko. *Startup bets chip set on ultrawideband home nets.* Electronic Engineering Times, June 24, 2002, p. 4.

provided more than enough success to interest even more users in attempting wireless communication. By about 1917, with increasing numbers of spark-gap transmitters attempting to transmit information over long distances, the result was pure bedlam. In an unsuccessful attempt simply to transmit over interfering signals, increasingly powerful transmitters and larger and larger antennas were tried. Transmit signal selectivity was initially determined only by the bandwidth of the antenna connected to the transmitter.[61] This was clearly not conducive to packing more users into the limited spectrum available, and circuit techniques to constrain the transmitted bandwidths even more, were developed during 1916-1919.

Through the introduction of narrowband, continuous-wave (CW) Morse Code Transmitters, the disruptive technology of narrow-band oscillators was also starting to have an impact. The British Marconi Company, however, felt that in its niche market the broader bandwidth inherent with a spark-gap transmitter was better to attract the attention of a radio operator aboard a nearby ship or ground station in the event of a catastrophe aboard a ship. CW transmitters, though, soon started breaking the communication distance records held by typical spark-gap transmitters. Meanwhile, the British Marconi Company stubbornly held on to its spark-gap transmitters, and, relative to the primitive state of lobbying Congressmen in that day, attempted to obtain protection against the replacement of its transmitters aboard ships-at-sea. Still, the writing was on the wall. Spark-gap transmitters could not be used in a crowded environment, and more and more signals were coming on the air each day. In just a few years, over 1918 through 1919, the British Marconi Company quickly lost the title of being the dominant wireless company to the new Radio Corporation of America (RCA) formed in 1919. The British Marconi Company had started to lose its market share due to disruptive technology changes.

The British Marconi Company's plan for survival was concentration in a niche marine market, where spark-gap transmitters still had an edge. Capitalizing on bandwidth-reducing technology, in 1924 the U.S. government attempted to further reduce the cacophony of transmissions through limiting both the transmitter power and the operating frequency of spark-gap transmitters. When the new regulations went into effect, spark-gap transmitters suddenly lost most of their communication competitiveness, nearly overnight. Still, the number of new transmitters, both spark gap and CW increased. In 1927, the U.S. government finally banned all spark-gap transmitters, even aboard ships far at sea. The result was that the British Marconi Company became more of an historical footnote than a continuing leader of the wireless industry.

Whereas radio signals started as broadband signals occupying very wide bandwidths, the trend for many decades was increasingly to decrease their bandwidths, approaching the minimum bandwidth necessary for transmission of information, while improving bandwidth efficiency. With the introduction of TM-UWB technology, the trade of bandwidth against power can again be re-evaluated, with the result that broadband signals once again represent a significant promise of creating new possibilities, and new opportunities for companies willing to invest in the new technology.

---

[61] This was also the case for the first generation UWB transmitters, developed in the 1980's.

Similarly, today's existing wireless companies risk becoming the modern-day equivalent of the British Marconi Company of 1927; surpassed by startup companies willing to develop TM-UWB products and market them. For existing companies, to fight the introduction of TM-UWB products through lobbying is very much like the attempts in the early 1920's to legislate the required use of spark-gap transmitters onboard ships to insure the maximum likelihood of attracting the attention of nearby radio operators in the event of a disaster. Legislation then only delayed the inevitable and the British Marconi Company fell in importance in 1927, whereas, from 1901 through 1919 it had been THE dominant wireless company. The lessons of history are often forgotten in technology circles, much to the financial detriment of companies, investors, and even individuals that forget the lessons. Instead of sending lobbyists to Washington to ban its use, wireless technology companies of today should instead become engaged in R&D to improve their understanding of UWB. The genie is already out of the bottle. It is better to profit from UWB technology than to try belatedly to ban its use. In the end, individual consumers, far-sighted companies, and the marketplace in general will be the ultimate winners.

2.0   <u>**PROPOSED UWB FOLLOW-ON RESEARCH ACTIVITIES**</u>

The focus of this project has been on UWB communication and UWB communication theory for application within future communication networks on Spaceports and Ranges. This work culminated in a New Technology Report describing a method to detect both cooperative and non-cooperative UWB transmitters, based on the CWT theory described earlier in this report.

The next step in understanding and benefiting from UWB technology is to extend this UWB research into through-wall and ground penetration radar applications, and especially into non-destructive inspection of non-metallic composites, to investigate the theoretical limits of UWB technology as applied to these allied areas.

The applications of such theoretical investigations should provide considerable benefit for the non-destructive inspection of materials anticipated for use on the existing ISS and on the planned OSP and NGLT launch vehicles, and on future payloads.

UWB technology has significant potential for providing key breakthroughs within both the communication networks of future Spaceports and Ranges, and for supporting the safe pre-flight processing of future launch vehicles and payloads.

3.0   RESEARCH CONTRIBUTORS

3.1   BIOGRAPHICAL THUMBNAIL SKETCHES

Gary L. Bastin, Ph.D.

Dr. Gary Bastin, Engineer Scientist, and Task Order Lead, was responsible for the detailed day-to-day execution of this project, including the preparation of the written final report, as well as of the choice and execution of the UWB evaluation hardware and test methods selected. He also derived and wrote the UWB theory sections of this document, and prepared the New Technology Report resulting from this research project.

Robert Chiodini

Mr. Robert Chiodini, Telecommunications Engineer, was responsible for the detailed testing of the UWB Evaluation Kit hardware. He wrote and executed the UWB test plan and test procedure, and collected and analyzed the test data.

PoTien Huang

Mr. PoTien Huang, as Principal Investigator, was responsible for monitoring the weekly progress of the overall project. Additionally, he supported the theoretical UWB investigations.

David A. Kruhm

Mr. David A. Kruhm was responsible for managing the overall UWB project.

4.0   <u>ACKNOWLEDGEMENTS</u>

Although there is always the risk of inadvertently forgetting someone, the UWB team nonetheless wishes to acknowledge especially the assistance and guidance provided by the following individuals, listed alphabetically. Without the continued support of these supporters who believed in the value of this project, this project could not have accomplished all its goals.

Name	Organization
Eric Denson	NASA-KSC
Temel Erdogan	ASRC-KSC
William G. Harris	ASRC-KSC
John Horan	ASRC-KSC
Chris Kerios	ASRC-KSC
Jules McNeff	NASA-HQ
Rich Nelson	NASA-KSC
Don Philp	ASRC-KSC
John Rush	NASA-HQ
Steve Schindler	NASA-KSC

5.0   __GLOSSARY__

Correlation   A mathematical method for determining the similarity between two functions or signal waveforms

Impulse Radio   An alternative term for UWB Radio

Monocycle   A single impulse transmitted by a UWB Transmitter; the fundamental unit of information transfer in UWB transmissions

Soliton   A solitary wave having both wavelike properties and particle properties, first observed as a wave traveling extremely long distances down a canal in the early part of the 19th Century. Its special properties are now used for transmitting 10 Gb/s and faster data rates over fiber optic cables over trans-oceanic distances, as well as for generating ultra wideband equivalent ranges of wavelengths in optical communication systems operating over fiber optic cable. Based on the work in this project, the waveform is found to have some key advantages for achieving ultra wideband spreading of electrical signals, such as for electronic warfare test equipment, over the use of monocycles.

UWB   Ultra Wideband, an adjective for indicating having either a fractional bandwidth greater than 25% or an occupied bandwidth greater than 500 MHz; now used as a noun, indicating UWB Radio